Fashionable Clothing from the Sears Catalogs

Late 1950s

Joy Shih

4880 Lower Valley Road, Atglen, PA 19310

To my brothers. I never told Mom those boots hurt your feet.

Thanks to Tammy Ward,
I couldn't have done it without you.
Thanks also to Desire Smith,
the expert in the field.

Library of Congress Catalog Card Number: 97-068110

ISBN: 0-7643-0339-2
Printed in Hong Kong

Published by Schiffer Publishing Ltd.
4880 Lower Valley Road
Atglen, PA 19310
Phone: (610) 593-1777; Fax: (610) 593-2002
E-mail: Schifferbk@aol.com
Please write for a free catalog.
This book may be purchased from the publisher.
Please include $3.95 for shipping.
Try your bookstore first.

We are interested in hearing from authors
with book ideas on related subjects.

Title page photo: Cover photo from Spring and Summer 1959, Minneapolis Edition 218, © Sears, Roebuck, and Co.

Sears Catalogs used with permission. Fall and Winter 1957, Atlanta Edition 215, Spring and Summer 1958, Seattle Edition 216, Fall and Winter 1958, Chicago Edition 217, Spring and Summer 1959, Minneapolis Edition 218, Fall and Winter 1959, Chicago Edition 219. © Sears, Roebuck and Co.

All items pictured in this book are from the late 1950s and are of interest to collectors and dealers of *wearable* vintage fashions. Highest prices are paid for items that are rare, or show originality of design and construction. For example, the shirtwaist dress, because it was so popular throughout the 1950s, has less market value than the chemise dress, which was popular for less than a year. Fabric also affects price. Higher prices are paid for pure silk than for acetate (which tends to fade, and even change color). Items which are extremely popular with vintage clothing collectors, such as gabardine suits and jackets for men and women, and rayon printed shirts and dresses, tend to bring high prices. Specialty clothing, such as nurses' uniforms, tend to be of greatest interest to costumers. Regional differences, trends in the market, and selling venue dramatically affect prices of vintage clothing and accessories. Highest prices are always paid for items in excellent condition. Prices given are for items in excellent condition. Similar items in poor condition may have little or no value. Current values are listed in brackets [ ] after 1950s prices.

Contents

upports

re Problems

r lower back

and sacro-iliac support
ll length of back has 2
. Pad is flannel covered,
ids. 2 back cluster lacers
you want. Garment of
overed boning. 4 garters.

rger than waist. Front
n. overall (11 in. waist
, 32, 34, 36, 38 in. State
Measure for Gales" on
nd 8 ounces.
. . . . $10.76

Back Support

as resulted from illness
ly good support. Now in
pport . . one for lumbar
ll boned. 2 removable
acers at each side give
ocaded cotton, tailored
Soft flannel protection
ters adjust. White.
0 in. larger than waist.
e "How to Measure" on

1 in. long; back 18 in. over-
, 28, 30, 32, 34, 36, 38, 40 in.
. . . . $12.72
in. long; back 15 in. over-
26, 28, 30, 32, 34, 36, 38 in.
. . . . $12.72

and Brace

eavier person.
pports permit
rain.

Introduction

Every American knows the Sears Catalog. In the 1950s, the Sears Catalog sold everything from motorcycles to vitamin pills, fur coats to kitchen sinks. Each generation has its own definition of what is fashionable clothing and for many, "fashionable" is determined by what the models wear in the Sears Catalog. The catalogs from the late 1950s, spanning the years between 1957-1959, provide a glimpse into what the average person wore during that era.

In this book, you are invited to look at the Sears catalogs through different eyes. Informative descriptions of each item give an idea into the kind of clothing styles popular during the period, color trends for each season, the "newest" types of fabric or fabric treatment to make life easier, the cost of the item in the 1950s, and if available, the current market values of the item as collectible clothing. It is interesting to note, also, the style in which the models are posed. What does this say about American culture in the 1950s?

About the Sears Catalog

In the Atlanta, Georgia Edition of the Sears Catalog for fall and winter of 1957, the catalog cover boasts the subtitle, "Serving all America's needs for the family, the home, the car, and the farm...at Sears everyday low prices." In 1957, more than twelve million customers throughout the United States used Sears catalogs as their buying guides. Sears offered the finest in fashion for the American family. In addition, the Sears Guarantee offered complete satisfaction with any product sold or your money back, so people could buy with confidence.

In the 1950s, Sears was willing to take your written order in any language, and offered a complete export service to over 100 countries around the world. For example, a customer could contact Sears of Philadelphia for orders to and from Greenland, Central America (except for Mexico), Caribbean Islands, South America, Europe, or Africa. For orders from or destined to Central Pacific Islands, Philippines, China, Japan, Korea, Australia, or New Zealand, the Sears at Los Angeles would handle the merchandise delivery.

A look at the specifics in a catalog tells us a lot about the times during late decade 1950s. Sales tax was collected if you lived in Alabama, Florida, Georgia, Mississippi, North Carolina, South Carolina, or Tennessee, and only at 3%. Telephone numbers used letters such as TR(inity)3-2311. The operator who took your order did not ask for a zip code after your address nor did customers have the convenience of using a 1-800 toll-free number to place orders. For returns, customers were instructed to box the item, wrap the box in strong paper, tie securely with string (something that will snag current postal equipment), and place a 3 cent stamp on an envelope pasted to the box. In the majority of cases, Sears customers were assured of delivery within 24 hours after receipt of their order by phone or by mail, but why couldn't you place an order to be delivered to Mexico?

Opposite page. An advertisement in the catalog showing how easy it is to order by phone, the "fastest and easiest way to shop." Fall/Winter 1958.

Catalog shopping was considered a convenient and efficient way to shop from home. It was also the only fashion source for women who lived in rural areas without access to a Sears store. The 1950s American woman used the Sears catalog as a fashion magazine, to keep track of the latest trends from Europe and the big cities. The telephone offered an easy way to keep the family well dressed and the home well equipped.

Sears Catalog as Fashion History

Fashion trends in the late 1950s will look familiar to those who were around at the time, and for those who were born later. Take a good look at the fashions and see the styles that have changed, and those that have not. A well-dressed woman wore a hat and gloves. Women's fashions accentuated the figure with tight waists and fitted bodice, and full skirts that make the hips look large. Imagine any woman today who would wear the "oval silhouette" styles that make the body look like a walking egg! Yet this was a popular style for suits and coats. Dress lengths remained firmly below the knee with the exception of formal gowns. Sweaters, however, look the same as the styles worn by today's women.

Men's fashions, on the other hand, appear about the same as today. Formal wear, suits, and shoes remain timeless fashion wear, spanning decades. Popular sportswear such as the *Jac-shirt*, a combination of a light jacket and a shirt, dates the item to the late 1950s. Hats, a staple for men at the time, have almost disappeared from current fashion trends with the exception of the baseball cap. In the 1950s, a baseball cap was worn only when playing baseball.

Only the car and perhaps the man's baggy slacks gives the clue that this photo is from the late 1950s. The woman's crewneck pullover and slim skirt, and the man's button cardigan are timeless fashions. Spring/Summer 1958.

Children's fashions of the 1950s appear formal to us today. At that time a distinction was made between school clothes and play clothes. Girls wore dresses to school and changed after school into playwear. Boys wore shirts and slacks and sometimes neckties to school, then changed to dungarees for play. Today there is little distinction between school clothes and play clothes, and many children do not even dress up for church.

Western wear in the 1950s was so prevalent in fashion wear that an entire chapter is devoted to showing this clothing. You will notice, however, that western wear was definitely not part of a woman's wardrobe except perhaps, in a tooled leather handbag. Children's cowboy and cowgirl outfits were worn not as Halloween dress-up costumes, but as everyday playwear. The Sears catalog sold many western wear outfits under the Roy Rogers brand name, after the popular children's television cowboy hero. Men's shirts also featured popular western styling.

Sears Catalog as Cultural History

A look through these pages will take you back to a time when life seemed relatively simple and people's roles were defined. Men went to work. Some women worked as nurses and in offices, or as waitresses and in other service jobs. Most women, however, stayed at home and planned the family's social life. Children went to school and came home to play. Everyone dressed up for everything, including sitting down to the evening's family meal.

You will note that all the catalog models were Caucasian. Men, women and children often pose with their hands in interesting positions. The photographers particularly liked to pose people with their hands cupped as if "calling" out to someone. Witness the many models in the following pages posing in the "calling" position.

Teen Sportswear. Bright barn red flannel shirt lined to match navy, black or beige jeans. Shirt, $2.44. [$20-25] Jeans, $3.94. [$40-45] Fall/Winter 1958.

Boys' Outerwear. Rain or shine cotton poplin pullover parka. Tan, red, or safety yellow. $3.77. [$45-50] Fall/Winter 1959.

Women's Fashions. On-the-job fashions for nurses were always a white blouse and skirt. Unlike other dresses, these outfits were advertised as "anti-static", "porous", and "lint-free." Caps are permanently starched. In Dacron taffeta, nylon taffeta, or cotton. $3.84-$10.84. [$25-30] Fall/Winter 1959.

Leisure time was definitely a woman's best friend. The number of stay-at-home dusters and leisure wear indicates that very few women worked outside the home. Posed sweetly in pouffy party dresses, drinking tea, talking on the phone, or poised in a suit ready to go shopping, all a woman needed was to look pretty and "feminine". The absence of active sportswear or athletic shoes says sports were not considered as legitimate leisure activities for the average woman. Even working women such as nurses looked starched and pretty in tight waisted uniforms, hardly comfortable for jumping on a gurney to resuscitate an emergency room patient.

Men's fashions show a dressy trend also. Imagine today's college men wearing three-piece corduroy suits around campus! A button down shirt, a tie and a sweater was considered casual wear outside of work. Outside of beachwear or poolside cabana outfits, men wore slacks, shirts, and shoes. Athletic shoes were reserved for actual sports acitivities.

In the late 1950s, hats were beginning to fade out as a fashion accessory, although some men still considered themselves not properly dressed without a hat. Even pajamas featured styles that looked formal. Male models were posed in "manly" settings, holding sporting trophies, sporting a pair of binoculars, hunting, measuring wood, or waiting for a commuter train. Leisure time was not at home in front of a television set, but at the beach or pool, hunting, fishing, or at a football game.

A bit of cultural history is evident by looking at children's Halloween costumes. Television program or movie characters such as Popeye or Superman were popular choices for costumes. It is interesting to see people's vision of robots in the 1950s by the image portrayed on a child's costume. Certainly, what children liked to wear for trick-or-treating defined pop-culture trends for the times.

Children's sleepwear, mostly boys' wear, favored "masculine" themes. The launching of Sputnik spawned many "space" designs, while sports and cowboy motifs continued to be popular.

Sears Catalog as Nostalgia

If you were around during the late 1950s, the fashions on these pages will certainly bring back memories. I recognized dress styles similar to ones I wore in elementary school. I was reminded that nylon stockings had a seam in the back of the leg and that a woman needed to wear a garter to hold them up. My dad wore a hat when he went out of the house. My brothers begged for cowboy boots with spurs. My mom thought it was cute to dress my younger sister and me in matching little sister-big sister outfits. I found the parasol purse that Dad bought for us one Easter. I remember how the bouffant petticoats under my dress used to make that swishing sound when I whirled around the room.

Today my children search through thrift shops to find clothing pictured on these pages. My college age daughter buys sweaters from today's fashionable clothing shops and they look exactly like the ones in the 1957 Sears Catalog. The only difference is it's probably not made in America and it costs a lot more. My adult son favors clothing from the 1950s, especially service workshirts and bowling shirts.

The fashionable clothing on the following pages are taken from actual Sears catalogs of the late 1950s. And when you know how much vintage clothing collectors are willing to pay for a dress you bought back in 1958, you'll be upset that you gave it to the Salvation Army. The lovely dress your mother gave away in 1957 only to have your child buy it back 40 years later could be here too! Have fun.

The author's brothers with those famous cowboy boots. 1958. *Author's collection.*

The Chiffon Evening Gown. Draped and pleated neckline accented by lighter tone. Long flowing panels, fully lined with acetate taffeta, nylon net underskirt. Nylon tricot chiffon. Shrimp. Full length skirt (45") $20.00. Short length skirt (33"), $19.00. [$75-95] Fall/Winter 1957.

Romantic Lace Dress "envisioned for the biggest moments of your life!" Fully lined with acetate taffeta, boned bodice, scallop-tiered skirt (about 33" long) has net underskirt. All-lace jacket is lined with net. Complete ensemble in nylon/acetate lace combined with nylon net. White. $30.00. [$85-90] **Sophisticated Lace Dress.** Cowl neckline of nylon tricot chiffon dips to V in back. Panels flow to hem, pointed bodice front and back, full circle skirt (about 33" long). Nylon and acetate lace over acetate taffeta. Light copen blue. $15.50. [$65-70] **Spring-o-lator Pump** with crystal ornament, 2 ¾" Lucite heel, leather sole. Clear vinyl plastic. $9.77. Fall/Winter 1957.

Women's Fashions

Formal Wear

Women's formal wear were often full skirted, frothy lacy gowns or dresses. Each season, a "dream dress" was presented in the catalog that was suitable as a wedding dress. Long gowns with flowing back panels were in fashion. Also popular were the many tiers of lace around a full skirt, puffed up with a net underskirt. Dress hems, if not long, remained consistently below the knee.

Semi-formal dresses are also shown here. Although some of the styles do not appear "formal," the fabrics of which they are made placed them in the formal wear section of the catalogs.

Color trends are evident for each season represented, but "feminine" pastels continued to be the colors of choice for formal wear. Black was not as popular as it is today. By the end of the decade, "far east colors" became the "hot colors." We might guess that the popularity of "The King and I" and "The Flower Drum Song" brought on the interest in the Orient. After all, American fashion trends usually followed those set by Hollywood star power.

Women's Formal Wear. Late Day Dress. "A very pretty, young-looking dress with flattering bustline pleats, bow-trimmed bands of taffeta, big billowing skirt." Red. $18.50. [$55-60] **Billowing Chiffon Late Day Dress.** Rayon chiffon, bodice lining edged with nylon lace, attached petticoat in acetate taffeta. Red. $15.50. [$55-60] **Imported Japanese White Beaded Bag.** Simulated pearls on organdy, rayon lined. $5.26. Fall/Winter 1957.

After Five Brocade Jacketed Dress. Jacket slit in back, bow trimmed, acetate taffeta lined. Slim moulded dress with softened bodice. Rayon and cotton brocade. Light Beige. $14.50. [$55-60] **Draped Satin Dress.** Acetate satin fabric, very fine rib. Softened bodice, moulded midriff, full skirt rounded by unpressed pleats. Nylon net petticoat, belt. Royal Blue. $17.00. [$45-50] Fall/Winter 1957.

Jacketed Dress. Cotton lace over rustling taffeta. Dress has deep scooped neckline in back, bow trimmed taffeta bands around Empire bodice and Jacket edge. Crinoline petticoat included. Light copen blue. $19.50. [$55-60] **Dream Dress.** "For the most wonderful moment of your life!" Nylon and acetate lace. Pleated nylon net insert in front of full gathered skirt, ruffled net underskirt. Fully lined with rayon and acetate taffeta. White. $26.50. [$85-90] **Timeless Lace Dress.** Nylon and acetate lace over rayon and acetate taffeta. Two pleated inserts of nylon net in skirt front. Collar ornaments and fabric belt. Dusty pink. $17.50. [$45-50] Spring/Summer 1958.

Panel Silhouette. Sheer yoke with scalloped cuff, attached cummerbund, skirt circled with scalloped tiers, attached sheer panels begin at waistline. Filmy silk organza, fully lined with rayon and acetate taffeta. Black. $19.70. [$65-75] **Flocked Flower Border Dress.** Scalloped scoop neckline and around skirt hemline. White nylon, bow-trimmed nylon velvet bands, dress lined with rayon and acetate taffeta. Black on white. $16.50. [$75-80] Spring/Summer 1958.

Pleated Lace Dress. Surplice bodice, taffeta cummerbund-effect. White nylon and acetate lace over pastel rayon and acetate taffeta. Light pink with white. $15.50. [$75-85] **Floating Panel Dress.** Shirred Empire bodice with embroidered daisies and beading on front midriff. Nylon tricot chiffon fully lined with rayon and acetate taffeta. Light blue. $19.50. [$75-85] Spring/Summer 1958.

Panel-Back Lace Sheath. Rayon/nylon lace over acetate taffeta, nylon tricot chiffon attached from neckline to harem-draped hemline, rayon velvet bows. Light peacock blue. $16.70. [$40-45] **Brocade Evening Slim Sheath.** Rayon/cotton brocade, jacket lined in acetate taffeta. Ivory white. $16.70. [$55-60] Fall/Winter 1958.

Chiffon Evening Gown. Nylon tricot chiffon, satin piping, waltz or long length. Raspberry. Waltz, $19.50. [$75-85] Long, $20.50. [$75-85] **Gala Scallop Tiered Dance Dress.** Flocked border design on filmy nylon, rayon velvet trim, bouffant skirt of nylon net. Pink with black. $17.70. [$95-125] Fall/Winter 1958.

Scallop Tiered Lace Dress. Acetate/nylon lace over acetate taffeta, sparkly trim on bodice, rayon satin midriff and bow. Mink beige. $19.50. [$50-55] **Chantilly-Type Lace Bouffant Dress.** Acetate/nylon lace, and acetate with peau de soie sheen. Frosted peach pink. $15.70. [$45-50] **The Dream Dress.** Nylon/acetate lace over acetate taffeta. Scalloped square neckline and butterfly bow at waist, pleated nylon net insert skirt, ruffled net underskirt. White. $26.50. [$85-90] Fall/Winter 1958.

Bell Silhouette Evening Dress with Stole. Rustling acetate taffeta, skirt lined with nylon marquisette. Aqua blue. $15.50. [$95-110] **Dotted Sheer Party Dress.** Duco print dots on sheer crisp nylon, acetate taffeta lined, satin circle bands trim waist and skirt. Yellow on white. $14.50. [$75-85] **Floral Bouffant Dress.** Sheer silk organza fully lined in acetate taffeta, nylon net underskirt, shirred midriff, matching bow and streamers. Turquoise. $15.50. [$65-75] Spring/Summer 1959.

Young Evening Look Dress. Filmy nylon tricot, fully lined with acetate taffeta, nylon lace underskirt, tricot top over glittering lace, shoulder ties. Party pink. $19.50. [$55-60] **Your Dream Dress.** Re-embroidered nylon and acetate lace with nylon net, lined with acetate taffeta. Tiny fabric-covered buttons down back, scalloped scoop neck, bouffant skirt with ruffled net underskirt. White. Waltz length, $32.50. Long length, $38.50. [$75-95] **Scallop Lace Tiered Dress.** Rayon and nylon lace over acetate taffeta, rayon satin midriff. Light dusty blue. $17.50. [$65-75] Spring/Summer 1959.

Harem-Skirted Moire Dress. Lustrous acetate and cotton moire dress, V front and back, big bow in front. Fully lined with net skirt interlining. Blue. $17.50. [$55-65] **Dream Lace Dress.** Nylon/acetate lace with nylon net. Scalloped skirt has apron effect in back revealing ruffled net underskirt. Back bow, set-in cummerbund. White. Short, $26.50. [$75-85] Long length, $29.50. [$75-85] **Lacy Evening Sheath and Jacket Ensemble.** Nylon/acetate lace over acetate taffeta. Slim sheath with scoop neck. Champagne beige. $18.50. [$55-65] Fall/Winter 1959.

Elegant formal evening wear. *From left.* **Billowy Nylon Tricot Formal Gown.** Fake pearl smocked bodice in front, V neck in back, fully lined. Pale green. $19.90.[$65-75] **Ruffly Tiers Bouffant Dance Dress.** Filmy nylon with an organdy texture, fully lined. Pale pink. $19.50. [$95-110] **Flocked Floral Bouffant Evening Dress.** Rayon matte jersey top and organdy-like nylon skirt,. Square cut neckline in back. Fully lined, net underskirt. Black and white. $19.50. [$85-95] **Enchanting Lace Dress.** Nylon and acetate lace over acetate taffeta, ruffled net underskirt. Rayon satin bands circle waist and skirt. Red. $15.00. [$65-75] Fall/Winter 1959.

This fashionable romantic brocade dress features a fur band around each sleeve. Curved neckline dips to a wide V in back. Cotton/rayon brocade, dyed mouton-processed lamb fur. Light powder blue. $13.84. [$75-85] Fall/Winter 1959.

Long-stemmed roses float on this dreamy formal strapless dress. Filmy nylon with the look and feel of organdy, fully lined in acetate taffeta, bordered with flocked roses. Harem skirt over nylon net. Pink. $16.50. [$95-110] His tuxedo is tropical rayon and Dacron, with a rayon satin shawl collar, rayon satin braid trim on pleated trousers. Blue-black. $39.50. Fall/Winter 1959.

Sophisticated semi-formal party dresses. *From left.* **Acetate Crepe Paisley.** Wrist zippers for snug fit, linen-look rayon overcuffs, V neckline in back. Nylon net petticoat included. Moss and teal green on deep blue. $14.84. [$30-35] **Porcelain Look Roses.** Embroidered roses on fine 100% pure wool jersey dress, fully lined in acetate taffeta. Scoop neck dips to V in back, cotton velveteen belt. Rose pink and moss green on white. $18.50. [$40-45] **Mock Bolero Front Blouse and Velveteen Toreador Skirt.** Cotton broadcloth blouse has back buttons, removable black nylon bow for washing. High rise skirt has rayon braid trim down sides. Blouse, $3.94. [$35-45] Skirt, $7.94. [$35-45] **Cloud of Chiffon Dress.** Sheer rayon chiffon over acetate taffeta. Billowy sleeves, finely tucked bodice front, all around midriff, detachable white silk organza overcollar and cuffs. Navy blue. $17.50. [$55-60] Fall/Winter 1959.

The Midriff Interest..in Arnel® Crepe, "looks and feels like precious silk, has a very fine rib." Scoop neck and long back zip. Bright blue. $14.00. [$55-60] **Slim Empire Dress with Chantilly-type Lace.** Nylon and acetate over acetate taffeta. Square neck, tiny cuff at front neckline. Flounce skirt back, acetate satin bow. Black. $14.50. [$95-110] Fall/Winter 1957.

Junior Formal Wear. Perfect for the winter formal dance, these ensembles were every girl's dream outfit. *From left.* **Princess Dress and Jacket.** Finely ribbed white faille (rayon and cotton) with blue lace flowers and glitter trim on jacket. Each lined in pale blue acetate satin. Set, $15.00 [$45-50]. **Bouffant Dress.** Sheer silk organza over acetate taffeta, accented with large bows. Flamingo pink. $16.50 [$60-65]. **Evening Coat.** Acetate satin flared black coat with white acetate satin lining. Self fabric buttons and bows. The interlining keeps the coat's flared look. $24.70. [$45-50] Fall/Winter 1959.

Draped Velvet. Crush resistant rayon velvet. Rhinestone ornament. Black. $18.50. [$75-85] **Pure Silk Print.** Very fine silk shantung. Contour-shaped back lined with cotton from midriff to below hips. Nylon net petticoat included. Amber print on black. $19.50. [$65-75] **Side-wrapped Satin.** Acetate satin. Wrap-around slim skirt with front unpressed pleats, back zip. Medium mink brown. $16.50. [55-60] Fall/Winter 1957.

Floral print acetate taffeta dress. Cut-out back, high boat neckline in front. Poppy red on bright rose. $15.70. [$85-95] Fall/Winter 1958.

Slender Embroidered Dress. Straight-line silhouette in linen-look rayon. Beige. $14.70. [$45-50] **Two-piece Straight Line Dress.** Overblouse top with cowl neckline and button trimmed yoke. Pleated skirt with all around yoke at hip. Navy. $16.50. [$35-40] Spring/Summer 1958.

Flower Print Low-Back Dress. Bateau front neckline, deep plunging back, all around midriff. Cotton. Sea turquoise blue print on white. $12.70. [$65-75] **Semi-formal Black Crepe Dress.** High draped yoke in front, back cut almost to waist with half belt and rhinestone buckle. Acetate and rayon crepe with very fine pebbled texture, lined in acetate taffeta. $13.70. [$55-65] Spring/Summer 1958.

Left. **The Harem Skirted Dress.** Skirt curves in and under in this rayon and accetate dress, deep V neckline in back, lined from midriff to hem. Autumn gold and brown on black. $13.84. [$65-75]
Right. **The Bell Silhouette.** Paisley print in acetate taffeta. "New tapered skirt is wide enough for walking." Peacock blue, green and black. $9.50. [$85-90] Fall/Winter 1958.

This elegant evening wear knit has an expensive hand-knit look.The top has scoop neck in back with open look stitched trim around neckline.The Full skirt has tiers of open loop stitching bound by bands of intricate close knit, elasticized waist. 100% Hi-bulk Virgin Orlon. Shell pink. $19.50. [$95-110] Spring/Summer 1959.

New Bell Silhouette. Pure silk surah, fully lined. Blue, aqua and lilac on bright blue. $18.50. [$85-90] Spring/Summer 1959.

Women's late-day dresses are featured in beautiful silk. **Billowy Bouffant Tile Print.** Silk organza. Delft blue, green, and white. $19.00. [$45-50] **Side Draped Dress.** Navy blue silk shantung with glittering side clasp. $12.84. [$55-60] **Paisley Print Blouson Dress.** Pure silk, collar tie can be worn as a sash, completely lined. Bright blue, turquoise and green. $13.96. [$45-50] Spring/Summer 1959.

Pure silk semi-formal wear. *From left.* **Slim Silk Shantung Sheath.** Silk chiffon sleeves and shoulders. Crushed, set-in cummerbund. Fully lined. Black. $15.50. [$95-110] **Silk Crepe Print.** Draped neckline has low-back bow. Royal blue and moss green on light green. $17.50. [$65-75] **Bell Silhouette in Silk Faille.** Unpressed front pleats shape the belled skirt. Emerald green. $18.50. [$55-60] Fall/Winter 1959.

Glamorous party dresses. *From left.* **Pure Wool Jersey Sheath.** Applique lace and embroidery. Neckline dips to a V in back. Fully lined. Turquoise blue. $14.50. [$50-55] **After-Five Night Dots.** Acetate taffeta flared skirt with detachable cummerbund. Neck dips to V in back. Extra belt included. Beige with black dots. $12.50. [$60-65] **Embossed Faille Jacketed Sheath.** Cotton and rayon, demi-jacket bowed in back, bow trimmed bodice on sleeveless dress. Ivory beige. $15.50. [$65-70] **Billowy Rayon Chiffon.** Set-in cummerbund. Acetate taffeta lined. Deep rose. $15.50. [$50-55] Fall/Winter 1959.

Dressier occasion Oriental styled dresses in the new "far east colors." *From left.*
Brocade Elegance.
Cotton/acetate brocade woven in a delicate pattern. Slim sheath with brief mandarin jacket. Jade green on antique gold. $17.50. [$95-125]
Pleated Classic Pleated Dress.
Arnel/rayon blend, bow trimmed belt buckles in back. Nugget gold. $11.54. [$85-90]
Obi Sheath.
Acetate/silk blend with wide attached obi sash in coordinating floral print. Bright blue. $12.84. [$95-110]
Fall/Winter 1959.

Women's Fashions

Day and Career Wear

Day wear for women in the late 1950s, whether it is for casual occasions or for work, is usually a dress. The fashions pictured on the following pages range from dressier versions to more casual ones, and include daytime party wear. It is interesting to note that when an outfit consists of two or more pieces meant to be worn together, it was still considered a dress.

The easiest place to shop for the latest fashions was in the Sears catalogs and the range of styles that Sears offered number in the hundreds. During this period, the "chemise" was an influential style that appeared in many variations. Europe was considered the center of fashion and many dresses boasted its French "styling." Full skirts remained a staple during late decade, with the shirtwaist dress being a popular choice.

For a better appreciation of seasonal and color trends, the day and office dresses are arranged in chronological order. These were the wonderful fashions to wear when going out.

Kerrybrooke Jumper and Blouse. Slim fitted acrilan/rayon jumper. Striped rayon blouse has button-look front but actually buttons in back, button-link French cuffs. "Wear it as a costume, or as separates paired with blouses, sweaters or skirts." Light brown jumper, brown and white striped blouse. $12.70. [$65-70] **Collared Dress.** Brushed flannel rayon/acetate, "It's a dress for travel, for office or classroom; perfect background for ingenious accessory changes." Dark gray. $7.74. [$55-60] **Fur-Look Trim Slim Dress** in brushed rayon/acetate flannel with fur-look acetate/rayon collar and cuffs repeated in pocket buttons. Light beige. $9.74. [$85-90] **Slim Striped Brushed Flannel Dress.** "Simply fashioned to lead a busy life..for campus, office, or travel." Rolled collar with white linen-look rayon undercollar. Rhinestone buckles on belt front. Medium dark gray. $6.54. [$40-45] Fall/Winter 1957.

Orlon Sweater Set. Beige. $9.74. [$55-65] **Slim Wool Flannel Skirt.** Seat-lined. Beige. $6.77. [$25-30] **Virgin Orlon Cardigan V-neck,** ¾ length sleeves. Copen blue. $5.74. [$40-45] **Cotton Broadcloth Shirt**. Convertible collar, ¾ length sleeves. Copen blue. $3.77. **Wool Tweed Flared Skirt** "looks luscious enough to be imported.." $8.71. [$25-30] Fall/Winter 1957.

Sleeveless Dress with Lacy Bulky Knit Jacket. 100% Worsted Wool Jersey. Amber and black. $19.50. [$35-45] **All Wool Hand Knit Look Sweater.** Mock turtleneck, raglan sleeves. Amber haze. $6.77. [$40-45] **All Wool Blanket Plaid Skirt.** Brushed mohair threads through stripes, leather-like plastic belt. Amber tones. $10.74. [$20-25] **Two-piece Blouson Dress.** Knit wool with bulky knit collar and cuffs, slim skirt lined in back. Amber. $17.50. [$35-45] Fall/Winter 1957.

Virgin Orlon Bow-tie Sweater "has three-fold charm." Detachable tie and neck tab to wear 3 ways..as shown with both tie and neck tab, with neck tab only or as open V-neck. Coral rose. $5.74. [$20-25] **Panel-flared Wool Flannel Skirt.** Self belt. Coral rose. $8.71. [$20-25] **Rib-collar Virgin Orlon Pullover.** Beige. $4.77. [$20-25] **Wool Tweed Straight Skirt**, seat-lined, arrow-stitched twin pleat detail. $7.74. [$35-40] **Bateau-neck Sweater in Ban-Lon®..Textured Nylon.** "New wonder yarn that resists piling or fuzzing..It's sleek, soft and shape retaining even after countless sudsings." Copen blue. $5.74. [$30-35] **Dutch-boy Wool Flannel Skirt**, snap-adjustable bow belt. Copen blue. $7.74. [$20-25] Fall/Winter 1957.

Dutch Boy Silhouette Ensemble. Arnel/rayon flannel. Rayon chiffon high midriff, scarf tied, crepe lined. Amethyst, deep and pale. $24.50. [$75-85] Fall/Winter 1957.

Rib-knit Jersey with Accordion Pleats. Orlon/wool. Amethyst. $19.50. [$65-75] **Pure Silk Blouse.** Petal collar, simulated jewel ornaments at cuffs. $7.97. **Straight-line Tweed Skirt.** Wool/silk blend. Dutch boy pleats, seat-lined in rayon/acetate taffeta. Amethyst and black tweed. $9.97. [$25-30] **Shaped Tunic.** Wool worsted crepe, twin collars with the top one of rayon satin. Acetate taffeta lined tunic; sheer nylon marquisette lined bodice. Amethyst. $21.50. [$75-85] Fall/Winter 1957.

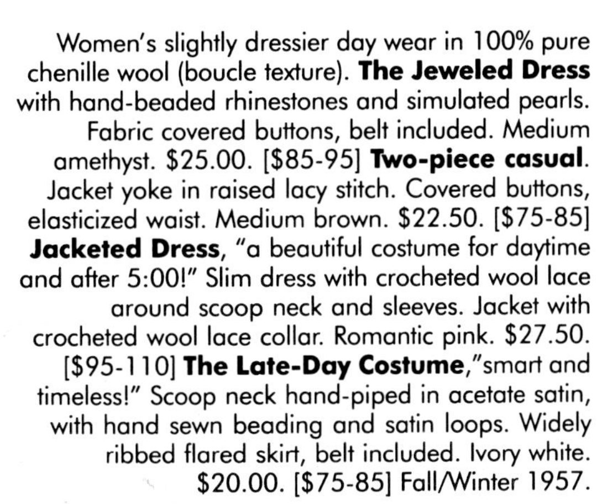

Women's slightly dressier day wear in 100% pure chenille wool (boucle texture). **The Jeweled Dress** with hand-beaded rhinestones and simulated pearls. Fabric covered buttons, belt included. Medium amethyst. $25.00. [$85-95] **Two-piece casual.** Jacket yoke in raised lacy stitch. Covered buttons, elasticized waist. Medium brown. $22.50. [$75-85] **Jacketed Dress**, "a beautiful costume for daytime and after 5:00!" Slim dress with crocheted wool lace around scoop neck and sleeves. Jacket with crocheted wool lace collar. Romantic pink. $27.50. [$95-110] **The Late-Day Costume**,"smart and timeless!" Scoop neck hand-piped in acetate satin, with hand sewn beading and satin loops. Widely ribbed flared skirt, belt included. Ivory white. $20.00. [$75-85] Fall/Winter 1957.

Printed Jersey Shirtwaist Dress. Orlon/wool blend. Hand bound buttonholes. Deep green print on bright turquoise blue. $19.50. [$35-40] **Wool Jersey Scarf Dress.** Fringed scarf drapes through shoulder loop. Without scarf, "the banded neckline is a perfect setting for jewelry." Fully lined with acetate taffeta, belt included. Red. $12.70. [$35-40] **Slim Woven Stripe Jersey Casual Dress.** Hip pockets, back kick pleat. Orlon/wool blend. Royal blue, with gray/black stripes. $14.50. [$55-60] Fall/Winter 1957.

The Important Black Dress. 100% wool jersey bodice, Chromspun acetate worsted broadcloth skirt with satin smooth back. Rhinestone ornament. Black. $10.74. [$65-70] **100% Acrilon® Jersey Pullover Dress.** "Relaxed easy fit dress you shape with the belt." Elasticized side gathers with no seam at waist. Ornament included. Bright copen blue. $9.54. [$45-50] **Silky Beaver Fur Felt Cloche**, wide grosgrain band with side tabs and rhinestone hoop. $7.97. **100% Nylon Jersey Dress.** Unpressed pleats all around skirt. Medium amber. $8.74. [$35-40] **Leopard-Look Beret** of rayon fur-like fabric. $2.83. **Orlon/Wool Blend Jersey Dress** with elasticized side gathers (no seams at waist), rib knit trim. Belt has draped cummerbund effect in front, buckles in back. Deep green. $10.74. [$45-50] Fall/Winter 1957.

Gabardine Slim Shirtwaist Dress, notched detail on collar and down. Metal buttons, silk square, fabric belt. Deep beige or Light blue. $10.00. [$75-85] Fall/Winter 1957.

The "chemise look" inspired by Paris featured shorter dresses fitted in front and loose in back. These Parisian styles were made for Sears in the USA. Spring/Summer 1958.

Black Crepe. Curved Cowl back panel, accented by bow, giving the dress a two-piece look. Textured rayon and acetate, fully lined in rayon organza. $16.50. [$45-50]

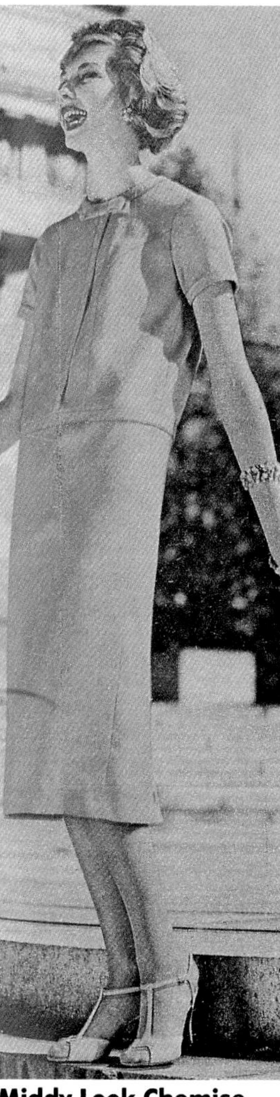

Ribbon Front Chemise. Screen printed acetate taffeta. Sun yellow and coral print with black rayon velvet ribbon. $12.00. [$40-45]

Middy Look Chemise. Linen-look rayon with silk nubs. Sea aqua blue. $5.74. [$30-35]

Belted Back Chemise. Hip placed half belt in back. Rayon and acetate straw cloth. Sun yellow. $8.74. [$40-45]

Houndstooth Coatdress. Woven checked cotton. Black and white. $7.54. [$55-60]

Bow-Trimmed Chemise. Acetate with satin-weave back. Sun coral. $14.00. [$30-35]

Bell-shaped "late day" fashion in silk shantung, fully lined in Pellon® (nylon, acetate, cotton) to retain its shape. Sea turquoise and blue print on white. $19.70. [$75-85] Spring/Summer 1958.

Straight and Narrow Big Plaid. Front belt effect repeats in back with a bow. Nubby textured homespun-look cotton, rayon and silk. Sea turquoise blue plaid. $12.70. [$65-70] Spring/Summer 1958.

Billowy Shirtwaist Dress. Silk and rayon. Sea blue and green print. $12.70. [$55-60] **Slim Jacketed Dress.** Wrinkle resistant rayon and cotton with the look of linen. Sleeveless dress with square front neck, cut low in back, dyed to match cotton lace. Sea aqua blue. Set, $16.50. [$75-80] Spring/Summer 1958.

Coin-Dot Party Dress. Fully lined bodice with square neckline and suspender effect in back, unpressed pleated skirt. Piazza cotton, an expensive shantung-textured wrinkle-resistant fabric. Rayon velvet sash. Red, pink and green on ivory white. $15.50. [$55-60] Spring/Summer 1958.

Big Collar Dress. Split collar in back, shoulder pleats for draping, unpressed pleated skirt. Shantung-textured cotton and rayon. Scattered dots on light green. $10.74. [$55-60] Spring/Summer 1958.

Two-Piece Polka Dot Dress. Surplice top with front waistband, pleated skirt. Arnel jersey. White with navy blue polka dots. 2-piece set, $12.70. [$85-90] Spring/Summer 1958.

Dotted Accessories. Hat, $3.97. [$35-40] Folded clutch bag, rayon lined, includes comb and mirror, $3.11. [$50-60] Reversible cape, white linen-look rayon and cotton polka dots, $2.97. Petal Ascot, ties 5 ways (illustrated guide included), $1.54. [$35-40] Umbrella with plastic tasseled handle, $4.77. [$45-50] Gloves, double woven nylon. $1.67. [$25-30] Spring/Summer 1958.

Low Back Summer Dress. "Dancing decolletage in back" with fabric-covered buttons. Cotton with white linen-look rayon edging. Sea turquoise blue and green print. $10.74. [$85-90] Spring/Summer 1958.

Romantic Chiffon Dress. Fine sheer cotton dress, tucking in front, billowy sheer sleeves, fully lined in acetate taffeta. Black. $12.70. [$75-80] **Cartwheel Hat.** Straw braid with velvet ribbon. $5.97. [$45-50] Spring/Summer 1958.

Big Sleeve Shirtwaist Dress. Deep V neckline, rayon satin banding on silk organza. Sea turquoise and blue on white. $23.50. [$75-80] Spring/Summer 1958.

Eyelet Embroidery Sleeved Blouse. White cotton batiste. $3.97. [$45-50] **Great Skirt with Half Crinoline.** Cinched with detachable rayon satin cummerbund. Arnel and cotton. Black with turquoise cummerbund. $6.97. [$35-40] Spring/Summer 1958.

Bubble Print Pleated Party Dress. Attached accordion pleated cummerbund. Dacron and combed cotton blend. Gold/sun coral/orange on white. $17.50. [$65-70] **Gloves.** Double woven nylon with scalloped edging. Sun coral. $1.97. Spring/Summer 1958.

Three-part Coordinates. Dacron/combed cotton blend, dyed to match braid and buttons on blouse, unpressed pleats on skirt, contrast cotton sateen cummerbund. Pink with bright rose. $13.00. [$55-60] Spring/Summer 1958.

Shirtwaist Dress. Shown in printed and solid styles. Roll-up sleeves with button-on tab. Light blue with navy patent-look belt. Sun gold, black on white with gold sash. $10.74. [$45-50] Spring/Summer 1958.

Cotton Nautical Dress. Long molded bodice with unpressed pleated full skirt, pearl-like decorative buttons all around. Navy blue with white braid trim. $9.74. [$45-50] Spring/Summer 1958.

Salty Air Coat Dress. Sailor collar with button-in bib, brass buttons. White/red and white stripe trim. $8.74. [$65-75] Spring/Summer 1958.

Long Sash Dancing Dress. Worn on or off shoulder. Cotton. Sun coral on sea turquoise blue. $4.97. [$95-110] **Imported Italian Beach Hat.** Hand woven straw braid with "lollipop" raffia dots. $3.57. [$20-25] **Tropical Fruit and Sea Shells Lei Necklace and Earrings.** Set $2.77. **Wicker Basket Bag from Spain.** Natural tan. $3.44. [$40-45] **Cotton Sail-cloth Play Shoes.** Clip-on fruit cluster sold separately. Sun coral. $2.83 pair. [$25-30] Clip-on cluster, $.47. Spring/Summer 1958.

Royal Blue and Black Silk Print Shirtwaist. Lined from waist to just below hips in acetate taffeta. $15.50. [$65-75] **Elegant Silk Crepe.** Detachable silk organza around neckline, acetate taffeta lined from waist to below hips. Black. $19.50. [$55-65] Spring/ Summer 1958.

Late Day Billowy Chiffon. Finely pleated top, fully lined with acetate taffeta attached petticoat. Rayon. Delft blue. $15.50. [$65-75] **Plaid Chiffon.** Sheer pima cotton, nylon net petticoat included. $14.70. [$55-60] Spring/Summer 1958.

Printed Collared Dress. Cotton broadcloth. Cotton velvet bow in front. $9.54. [$55-60] **Two-Piece Casual Outfit.** Top with band collar, rick-rack trim. Cotton satin. Black and white print. $9.74. [$75-80] **Sun Princess Dress.** Nylon scarf can be worn different ways. Scarf shown pulled through looping and draped over shoulders. Embossed cotton sunflower print. Sun, gold, and orange print. $6.74. [$75-85] **Full Skirted Silhouette Dress.** Dyed to match lace, deep V back accented with fabric bows. Tucked front and back bodice. Cotton broadcloth. Light blue. $11.70. [$45-50] **Sea Color Cotton Print.** Big tab-trimmed patch pockets, fabric belt. Sea blue and green. $6.74. [$45-50] Spring/Summer 1958.

"Newest" Blouson Two-Piece Dress. Sheer all wool worsted crepe. Detachable dickey of cotton pique. Sapphire Blue. $17.50. [$95-110] Fall/Winter 1958.

Trapeze Silhouette Rayon Plaid Jacket and Skirt. Overlapping box pleats in skirt. Light red and autumn tan plaid. $18.50. [$65-70] Fall/Winter 1958.

Two-Piece Chemise. Button-back overblouse with leather-look trim. Black with white tweed. $14.50. [$55-65] Fall/Winter 1958.

Flouncy Floral Chemise. Collar-less neckline dips to V in back. Acetate taffeta. Red roses on black. $14.00. [$75-85] Fall/Winter 1958.

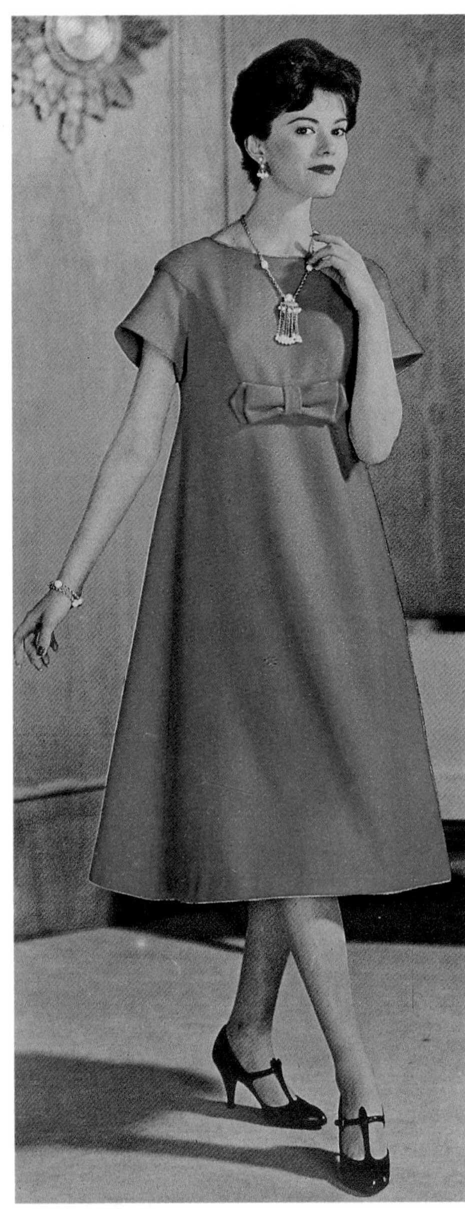

Paris-inspired Trapeze Dress. 100% wool flannel. Curved yoke with one big button at back neckline. Back lined. Turquoise blue. $8.74. [$65-70] Fall/Winter 1958.

Big Collar Chemise Coat Dress. Back of bodice has bow-trimmed inverted pleat. Coral red. $14.84. [$65-75] Fall/Winter 1958.

Glen Plaid Dress. Arnel and cotton. Brown and white. $10.74. [$40-45] Fall/Winter 1958.

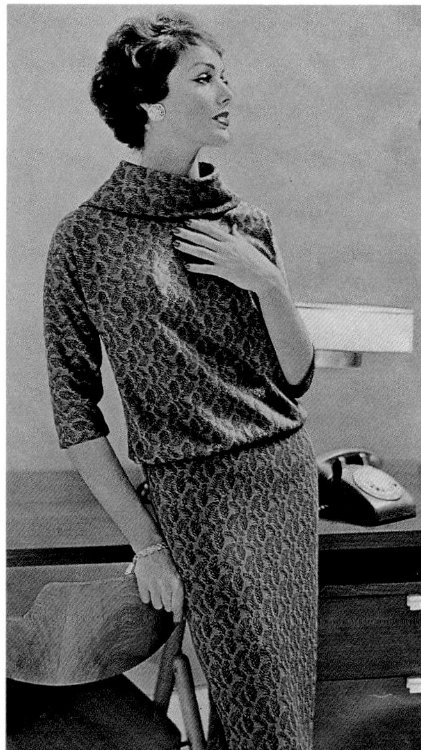

Leaf Patterned Bloused Cotton Knit. Two-piece dress, cowl collar, elasticized band at waist. Olive and black. $8.74. [$55-60] Fall/Winter 1958.

Honeycombed Textured Cotton Knit Chemise Dress. High rise waist double-breasted look, fabric flower. Sapphire blue. $13.84. [$75-80] Fall/Winter 1958.

Basket-Type Weave One-Piece Chemise. Two-piece look, fully lined, wool/ angora Rabbit hair/nylon. Deep moss green. $21.50. [$45-50] Fall/Winter 1958.

Corduroy Coat Dress. Button-on adjustable belt high in front, draped loosely in back. High and low wale corduroy. Autumn rust. $7.54. [$65-70] Fall/Winter 1958.

Wool Flannel Chemise Jumper. D-ring adjustable tabs. Autumn green. $11.97. **Paisley Cotton Broadcloth Blouse.** $3.97. [$45-50] Fall/Winter 1958.

Leaf Patterned Chemise Dress. Gathered and bowed in front, buttons to below waist in back. Red and autumn tan. $7.54. [$45-50] **Two-Piece Loose Fit Cotton Knit Chemise Set.** Sapphire Blue and black plaid. $8.97. [$75-85] **Assorted Acessories.** [$20-35] Fall/Winter 1958.

The Decollete Late Day Dress. Pure wool crepe with acetate satin. Black. $15.70. [$65-70] **Jacket, Chemise Jumper and Blouse.** Wool flannel and combed cotton broadcloth. Ruby Red. Set, $16.70. [$45-50] **Bloused Casual Dress.** Wool woven plaid, lined bodice buttons in back. Red/black. $9.54. [$45-50] **Drawstring Jacketed Dress.** Wool flannel with fur fabric Orlon collar. Deep emerald green. Set, $14.84. [$45-50] **Shirtwaist Chemise.** Wool flannel, pleats in back from yoke to buckled half belt at waist, detachable rayon collar. Camel tan. $13.84. [$45-50] Fall/Winter 1958.

"Waistlines are up...the rise of a new fashion." High waisted, under a high bodice, and a tapered skirt to below the knee was the sophisticated look of the season. *From left.* **Floral Printed Cotton Hopsack Dress.** Gathered loosely in back and front with a bow. Blue/green on aqua. $9.84. [$45-50] **Sophisticated Black Crepe.** Shirred drawstring effect, acetate/rayon crepe. $12.80. [$50-55] **High Belted Dress.** Detachable dickey, inverted pleat in back. Ivory beige. $6.84. [$65-70] Spring/Summer 1959.

Two-piece Flare Dress. Top buttons in back. Combed cotton with textured surface. $10.84. [$45-50] **Imported Pure Linen Shirtwaist.** Fringed collar and neck trim, dress lined to hips. $13.00. [$60-65]
Spring/Summer 1959.

Hi-Line Dotted Dress. Rayon/silk textured fabric, belted high on waist. $10.84. [$55-60] **Floral Bouffant Cotton Dress.** Wide cummerbund. $7.96. [$50-55]
Spring/Summer 1959.

Fresh and bright spring green in varied shades were new fashion colors for spring 1959. Shown here are a rayon and silk floral ensemble with a high waist jacket, and a dacron/cotton blend embossed scallop design dress. Ensemble, $17.00. [$40-45] Scallop Dress, $10.54. [$40-45]
Spring/Summer 1959.

The shirtwaist dress in hip belted, classic waist belted, and high belted styles. In Arnel/cotton **Gold Paisley Print**, $10.84 [$35-40], combed cotton **Multicolor Pastel Plaid**, $8.66 [$35-40], and textured rayon and acetate **Dusty Green Slim Dress**, $9.77. [$55-60] Spring/Summer 1959.

Geranium Rose was a new color for spring 1959, perfect for a garden party. *From left.* **Draped Bodice Dotted Dress** in sheer Dacron with overpanels in front. $12.90 [$55-60]. **Slim Jacketed Lacy Dress.** Nubby textured rayon and silk. $16.50. [$65-70] **Floral Geranim Print Halter Dress.** Cotton with cotton velveteen green belt. $7.88. [$75-80] Spring/Summer 1959.

Gingham Check Contour Belted Two-Piece Dress. Leaf green and white. Set, $6.84. [$35-40] **Woven Check Jacketed Dress.** Rayon and acetate check fabric, pull-through belt on jacket. Blue and white checks. Set, $8.96. [$35-40] **Houndstooth Check Shirtwaist.** Cotton voile with tucked front. Black and white. $11.84. [$50-55] **Embroidered Camisole Top Check Dress.** Cotton/Arnel in red and white. $6.84. [$50-55] Spring/Summer 1959.

Three-piece WovenCheck Outfit . Arnel/cotton top with a white overcollar and colorful bands around the waist, a box pleated check skirt, and an additional slim skirt in cotton broadcloth. Black and white with red. Set, $10.90. [$35-45] Spring/Summer 1959.

Dress with Beaded Orlon Sweater. Rayon/dacron, sweater hand beaded. Geranium Rose. $16.50. [$65-70] **Rib Knit Chenille Orlon Dress.** Nubby textured with belt to wear empire fashion or tied classically at waist. $21.50. [$75-80] Spring/Summer 1959.

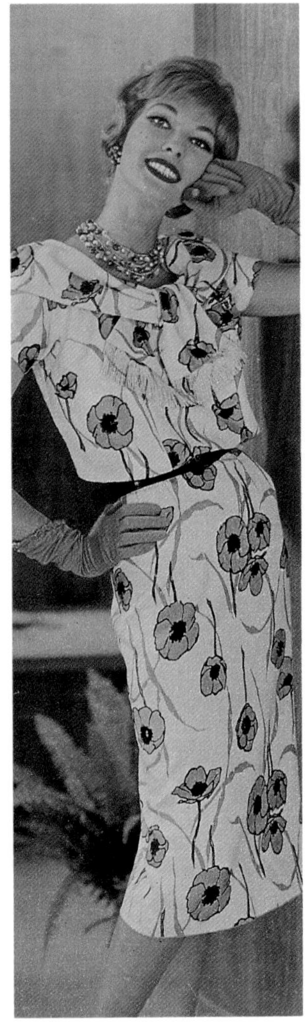

Brown was the "New Neutral" for spring and summer. The new rich tones were called Spring Brown but came in many mid-range tones. **Flowering Poppies Blouson Dress.** Expensive textured cotton that "looks like fine shantung." Fully lined with shawl-tied fringed collar detail. Mocha and gray with green on light beige. $15.50. [$60-65] Spring/Summer 1959.

Two-Tone Plaid Full Skirted Dress. Dacron/cotton, contrast piping, wide box pleats. Spring brown on light beige. $12.66. [$50-55] **Raised Waistline Dress.** Shantung-textured cotton/rayon with white button trim. Spring brown. $7.84. [$60-65] Spring/Summer 1959.

The newest styles in textured weave cottons are featured in fitted and high waisted designs. *From left.* **Sheath and Button Back Plaid Jacket.** Basketweave hopsacking 100% cotton with high back belt. $7.54. [$45-50] **Plaid Sheath with High Bodice Band.** Cotton hopsacking basketweave. Spring brown and orange with white highlights. $5.84. [$50-55] **Bright Multicolor Stripe Free-Swing Jacketed Dress.** Ivory dress with multicolor detailing. $8.50. [$50-55] **Flowering Sheath.** Square neckline accented with a large bow. Spring green, blue and brown on white. $9.80. [$40-45] **Hi-Line Basketweave Dress.** Big collars, high belted. Spring brown. $7.84. [$50-55] Spring/Summer 1959.

Linen-Look Rayon Jacketed Dress. Square neck sheath with finely tucked front belted jacket. $8.90. [$60-65] Spring/Summer 1959.

Dress with Slim Coat. Relaxed sheath tied at waist in front, loose in back. Double breasted plaid rayon coat, lined in front with acetate taffeta. Spring brown dress, gray, brown and black plaid coat. Each $15.70. [$65-70] Spring/ Summer 1959.

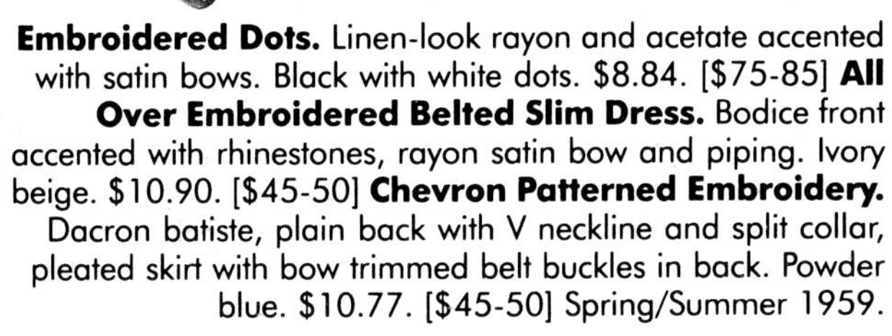

Embroidered Dots. Linen-look rayon and acetate accented with satin bows. Black with white dots. $8.84. [$75-85] **All Over Embroidered Belted Slim Dress.** Bodice front accented with rhinestones, rayon satin bow and piping. Ivory beige. $10.90. [$45-50] **Chevron Patterned Embroidery.** Dacron batiste, plain back with V neckline and split collar, pleated skirt with bow trimmed belt buckles in back. Powder blue. $10.77. [$45-50] Spring/Summer 1959.

Daytime dress with fancy embroidery. **Deep-Toned Embroidered High Waist Dress.** All over pattern on lighter color cotton satin. Rose pink on light pink. $12.80. [$55-60] Spring/ Summer 1959.

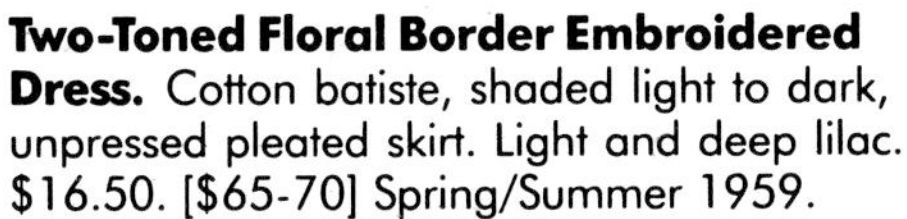

Two-Toned Floral Border Embroidered Dress. Cotton batiste, shaded light to dark, unpressed pleated skirt. Light and deep lilac. $16.50. [$65-70] Spring/Summer 1959.

Cotton dresses in Shagbark knotted and satin-smooth weaves. **Deep-Toned Double Breasted Plaid.** Rounded lapels and cuffs. Cornflower medium blue.$8.84. [$45-50] **Chevron Pattern Slim Shirtwaist.** Tiny buttons and tab trim. Spring green. $7.74. [$55-60] **Soft-Toned Plaid Jacket and Dress.** White pique trim. Tan/gray/white plaid. $10.90. [$55-60] **Tiny Textured Pincheck.** Criss-cross tab trim, knotted weave Shagbark fabric. Geranium Rose. $10.84. [$35-40] **Textured Stripe Shirtwaist.** Shagbark fabric combed cotton with knotted weave. Convertible collar and roll up sleeves with button tab. Gold, tan and coral. $10.77. [$50-55] Spring/Summer 1959.

"It's here! The dress you can machine wash, tumble-dry and wear!" **Full Circle Pleated Dress.** Dacron/nylon/cotton, tucked front bodice with button trimmed belt effect. Geranium Red. $16.50. [$45-50] **Slim Blouson Dress.** Dacron/cotton, lined bodice, button trim. Gold and black on white. $9.84. [$35-40] **Portrait Collar Dress.** Dacron/cotton broadcloth, button trimmed over-panel on bodice front, pull-through belt, unpressed pleated skirt. Deep copen blue. $10.84. [$45-50] **Tucks and Tiny Looping Shirtwaist.** Dacron/cotton, with tucks on front of bodice and all around the full skirt. Apricot with white looping. $12.84. [$50-55] Spring/Summer 1959.

Stylized print in rayon and silk surah that looks like pure silk. Pointed yoke at waist in front, half belt in back. Turquoise blue, green and gold on white. $11.96. [$85-95] Spring/Summer 1959.

Drop Waist Pleated Skirt. Cuffed hip band and tiny knife pleats. Aqua blue. $11.84. [$60-65] Spring/ Summer 1959.

Lace and Tucks Woven Stripe Dress. Button trim on bodice front, contrasting bow trim belt. Black and white. $7.84. [$60-65] **Dyed to Match Lace Dress.** Shirred sleeves, lace edging all around bateau neckline. Light blue. $6.84. [$45-50] **Twin Pleated Floral Sheath.** Cotton voile top misted over cotton satin sheath. Geranium rose/maize/white. Two-piece, $7.97. [$55-60] Spring/Summer 1959.

Embroidered Panel Dress. Sheer cotton. Light pink /white. $11.84. [$65-75] **Bare Shoulder Dress.** Cotton lawn shadowy printed plaid, bordered by double bands of white organdy. Lace medallions around border and on bodice front. Black and white. $7.54. [$65-75] **Harem Skirt Floral Dress.** Bateau neckline dips to V in back, skirt completely lined to retain shape. Geranium rose and gold print on white. $6.84. [$45-50] **Drop Waist Dress.** Hipline has band of white cotton lace, ribboned and bowed in nylon velvet. Combed cotton broadcloth. Mint green. $6.84. [$50-55] **Printed Jacketed Sun Dress.** All around pleated midriff. Turquoise blue, green and white print. $10.84. [$35-40] Spring/Summer 1959.

Continuing in the Oriental flavor, these dresses were advertised as "far east colors." **Full Skirted Side Button Dress.** Shantung-textured cotton and rayon. Bright turquoise green. $8.84. [$85-90] **Siamese Stripes.** "Vibrant muted tones that reflect the splendor of the Orient." Arnel/rayon broadcloth with silk-like luster. Multi-color. $15.84. [$65-70] Fall/Winter 1959.

Classic Silk Shirtwaist. Cluster-pleated full skirt, in solid and printed styles. Royal blue or blue on peacock. $18.50 each. [$40-50] Fall/Winter 1959.

Knit dresses in fine worsted wool and machine washable Orlon. **Mock Pearls and Braid Two-piece Dress.** Wool. Sapphire blue. $13.84. [$85-90] **Narrow Rib Knit Shirtwaist.** Wool. Spice brown. $13.50. [$75-80] Fall/Winter 1959.

Pebbly Textured Knit Set. Orlon. Red.$10.50. [$65-70] **Three-piece Lacy Knit.** Orlon. White. $13.50. [$75-85] Fall/Winter 1959.

Savory Plaid Shirtwaist. Rayon/wool blend updated with stitched center pleats. Paprika, gold and tan. $10.84. [$45-50] **Striped Paisley Slim Sheath.** Surplice neckline, shoulder tucks. Rosewood tan, black on old gold. $9.84. [$40-45] **Casual Tailored Sheath.** Buttons to below waist. 50/50 wool/rayon boucle textured fabric. Poppy red with black. $13.84. [$40-45] Fall/Winter 1959.

Women's daywear in new fall "spice colors." **Textured Stripe Knotted Weave Dress.** Galey and Lord original combed cotton Shagbark. Interesting shaped bodice. Light spice tan/gold/coral, with coral belt and scarf. $10.84. [$50-55] **Basic Jersey Sheath.** Orlon/wool blend, semi-lined dress. Paprika. $8.84. [$85-95] **Two-Tone Iridescent Dress.** Rayon/acetate shantung textured fabric. Rounded collar ties in bow, full circular skirt. Green and spice brown. $8.84. [$50-55] Fall/Winter 1959.

Blouson Wool Crepe. Rayon satin bow trim, semi-lined. Teal blue. $16.50. [$50-55] **Sheath and Matching Stole.** Checked wool jersey with herringbone weave. Button trimmed pockets, lined from waist to below hips. Blue, olive and violet. $18.50. [$65-70] **Rich Wool Flannel Jacketed Dress.** Jacket banded all around, bowed in front. Fully lined sheath. Cranberry red. $14.50. [$50-55] Fall/Winter 1959.

Women's Fashions

Everyday Dresses

In the 1950s, most women did not work outside the home. Everyday dresses were fashionable, inexpensive, and affordable stay-at-home dresses. Accessorized with hat and gloves, these dresses were suitable to wear outside for shopping errands, doctor's appointments, or lunch dates. What makes them everyday dresses is the price. In the late 1950s, an everyday dress from the Sears catalog could be purchased for as little as $2.77.

Assorted cotton dresses. **Bright Rose Flower Print on black**. Shoulder pleats, flared 16-gore skirt. $3.76. [$50-55] **Oriental Print**. Amber/brown print on white. $3.76. [$45-50] **Cape Collared Tie-print Coat Dress**. Gold/white print on turquoise blue. $3.76. [$45-50] **Modern Abstract Print Dress**. Amber nylon scarf pulls through loop at neck. Amber with black print. $3.76. [$55-60] Fall/Winter 1957.

Tie-Print Coat Dress. Medium blue with beige/black print. $3.74. [$35-40] **Button and Bows Broadcloth**. Box-pleated skirt. Rose-red with navy tab and bows. $3.74. [$45-50] **Woven Check Gingham Coat Dress**. Button front, fabric belt. Black and white checks. $3.76 [$55-60] **Graceful Printed Dress**. Black cotton lace design on collar, 16-gore skirt with black piping accents. Gold with red and black print. $3.76. [$60-65] Fall/Winter 1957.

Kerrybrooke Plaid Gingham. Button tab collar. Amber plaid. $5.74. [$40-45] **Duco Dot Broadcloth Slim Dress.** Silk square. Black with bright blue dots. $5.74. [$65-70] **Striped Coat Dress with Back Yoke.** Medium blue, gray, and black woven stripes. $6.74. [$35-40] Fall/Winter 1957.

Tweed-look Print. Medium gray with rose and white. $2.77. [$55-60] **Tattersall Checked.** Decorative buttons on collar and sleeves. Red/gray on white. $2.77. [$40-45] **Plaid Printed Coat Dress.** High round cardigan neck front and back. Lilac/aqua/gray plaid. $2.77. [$45-50] **Rosebud Pattern Dress.** Pink embossed cotton trim. Pink/ medium blue on black. $2.77. [$85-90] Fall/Winter 1957.

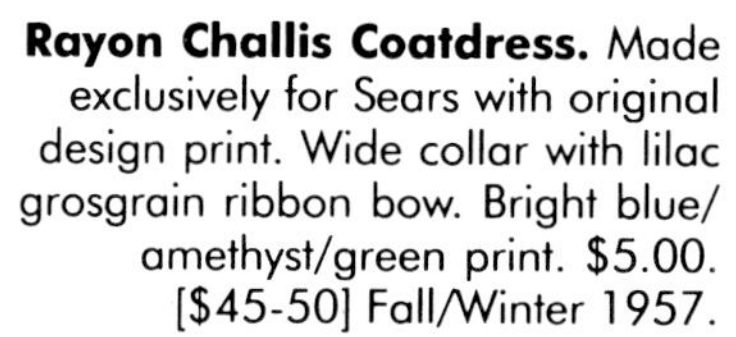

Rayon Challis Coatdress. Made exclusively for Sears with original design print. Wide collar with lilac grosgrain ribbon bow. Bright blue/ amethyst/green print. $5.00. [$45-50] Fall/Winter 1957.

Kerrybrooke Two-piece Dress. Shirtwaist top with roll-up sleeves, Parisian border print skirt. Gold top, gray with gold/aqua print skirt. $5.74. [$75-80] **Wrap-style Dress.** Surplice neckline, front-wrap skirt with belt. Medium amethyst. $6.74. [$60-65] **Surplice Everglaze® Cotton Print Dress.** Midriff gathers has all-around cummerbund effect. Red/gold. $5.74. [$55-60] Fall/Winter 1957.

Misses cotton everyday dresses "for sport or casual wear." These styles feature bows, and belted full skirts. $5.74 each. [$40-45] Spring/Summer 1958.

Budget-minded women's cotton dresses. **Ivy League Striped Tailored Classic.** Deep roll-up cuffs. Red, black/white stripes. $2.77. [$35-45] **Polished Chambray Step-in Dress.** Embossed cotton edged with loops trim on rounded yoke and pockets. $2.77. [$35-40] **"Dancing Doll" Print.** Mandarin collar piped in black, black decorative "safety pins", full gathered skirt. Bright red with multicolor print. $2.77. [$45-50] **Checked Print Coat Dress,** with "new" cone-shaped pockets. Black/white checks. $2.77. [$35-40] Fall/Winter 1957.

Misses everyday cotton dresses. $4.76 each.[$45-50] Spring/Summer 1958.

Misses everyday cotton dress featuring polka dots and floral prints. Styles include slim styles, shirtwaist, and coat dresses. In textured rayon and rayon challis. $5.00 each. [$45-55] Spring/ Summer 1958.

Misses everyday economically priced cotton dresses. $2.77 each. [$35-45] Spring/Summer 1958.

Women's Kerrybrooke cotton and cotton knit every day dresses. Hems are consistently placed below the knee. $5.77-$7.54. [$45-55] Fall/Winter 1958.

An assortment of Dan River Cotton tailored dresses at an inexpensive $5.84 [$40-50] each. *Top left and clockwise.* **Fancy Front Schiffli Embroidered Dress.** Light copen blue. **Tab Collared Woven Gingham Plaid.** Gold, black and white. **Buttons and Bows Leno-weave Gingham Coat Dress.** Aqua blue woven plaid. **Easy-Fitting Nubby Textured Woven Plaid.** Geranium pink and white. Spring/Summer 1959.

Everyday dresses in Arnel, Dacron, or nylon taffeta. The shirtwaist dress was still a staple in fashion. *Third from left.* Worn without a belt, this green dress style was considered an "eased" shirtwaist. All priced at $5.84. [$45-55] Spring/Summer 1959.

Budget-minded cotton dresses for everyday wear. Note the popularity of the tied-at-the-neck scarf and tied-at-the-waist sash. All dresses at the unbelievable price of $3.84. [$50-65] Spring/Summer 1959.

Sears' "lowest price washable cottons", affordable to most, were priced at $2.77 each. [$35-45] Spring/Summer 1959.

Affordable cotton dresses for everyday wear in several fitted-at-the-waist styles. Priced at only $5.84 each. [$55-65] Spring/Summer 1959.

Advertised as all-day cottons for the busy homemaker. Apparently stay-at-home women wore heels every day. All priced at $3.84. [$45-50] Fall/Winter 1959.

Casual dresses at only $5.84 each. [$55-60] Assorted styles and fabrics, suitable for every occasion. Fall/ Winter 1959.

There's no excuse not to be well dressed when every woman can afford these washable cotton everyday dresses priced at an unbelievable $2.77 each. [$40-45] Fall/Winter 1959.

Women's Fashions

Casual Separates - Sets

Separates are blouses, sweaters, slacks, skirts, or jumpers to mix and match. If purchased in a set, each piece may be worn separately with pieces from other outfits. Separates were favored by college coeds and adults alike because they offered women many creative options.

In this section, casual separates from the Sears catalogs are shown in coordinated sets, or two or more "separates" placed together to give the customer an idea on how to create an interesting wardrobe.

WASHCORD® corduroy for campus or career. **Two-piece Suit.** Button front jacket fully lined with cotton broadcloth print to match blouse. Medium Blue. $14.74. [$55-60] **Sleeveless Print Blouse** in washable cotton broadcloth. Medium blue, black and white novelty stripe. $2.83. [$15-20] **Flare Jumper with Gold-color Accents.** Matching belt. Black. $8.97. [$45-50] **White Scarf-tied** Blouse in washable Acrilan Jersey. $3.27. [$15-20] **Slim Jumper** "boasts a beautifully styled bodice and fashion's newest pet..soft front skirt folds." Red. $6.97 [$45-50]. **White Ruffly Cotton Broadcloth Blouse.** Back buttons. $3.77 [$15-20]]. **Mock Monogram Blouse.** Washable Acrilan Jersey, back zip. Light blue with blue monogram and trim. $3.77 [$20-25]. **10-Gore Wide-Flare Skirt.** Medium blue. $6.77. [$45-50] **Chic Roll-Sleeve Print Shirt in Everglaze® Cotton** "looks like expensive tie-print silk." Wear shirttails in or out. Cadet blue on gold. $3.77. [$15-20] **Frontier-style Corduroy Pants.** Slightly tapered, belted to match print shirt. Black. $4.77. [$20-25] Fall/Winter 1957.

Two-piece sweater sets for Juniors. **Kerrybrooke Shawl Collar Set.** 100% wool jersey blouse trimmed with surface-brushed flannel skirt fabric. Fringed-edged tied collar, leather-belted skirt "a generous half circle of fullness." Turquoise top with gray/turquoise/white box plaid skirt. $15.67. [$50-55] Skirt only, $9.97. [$25-30] **Hip Pocket Set.** Woven tweed with jersey, seat-lined skirt. Gold with gold tweed. $12.72. [$40-45] Skirt only, $7.44. [$20-25] **Scarf-Collared Set.** Wool Jersey blouse, leather-belted woven plaid pleated skirt. Cocoa top with blue/white/cocoa plaid skirt. $15.67. [$40-45] Skirt only, $9.97. [$20-25] **Back-Wrap Set.** "newest fashion to sweep the nation." Wool jersey top. Woven tweed back wrap skirt, fully lined, buttons on back. Black top with black and white tweed skirt. $13.64. [$50-55] Skirt only, $8.71. [$20-25] **Convertible Neckline Set** "changes its looks by buttoning tab on or off." 100% Wool jersey blouse. Wool tweed/nylon blend skirt. Coral rose with coral tweed skirt. $13.44. [$50-55] Skirt only, $8.71. [$20-25] Fall/Winter 1957.

Blazer for Her. Pure wool with imported Indian emblem and brass coin buttons, fully lined in rayon. White. $17.90. [$20-25] **Sport Coat for Him.** Textured 95% wool/5% nylon, three-button style. Dark blue. $24.50. [$50-55] Fall/Winter 1959.

Two-Piece Chemise Set. Overblouse and plaid shirt, removable plaid tie. Wool jersey. $15.97set. [$50-55] **Side Button Coat Dress.** Woven plaid 100% worsted wool in a twill weave. Red plaid. $17.70. [$45-50] **Pleated Plaid Skirt.** Wool and nylon cashmere blend. Rust and green plaid. $10.97. [$20-25] Fall/Winter 1958.

Campus casuals for the coed. **Novelty Knit Checkerboard Sweater Jacket.** Rack stitched mandarin neckline and front panel, 3/4" sleeves, 100% Virgin Hi-bulk Turbo Orlon. White. $9.74. [$15-20] **Box Pleated Houndstooth Check Skirt.** Wool/nylon/rayon blend. Black and white. $8.90. [$15-20] **V-neck Vest.** Lamb's wool/angora rabbit hair/nylon. $6.97. **Slender Skirt.** Side front pleats with button detail, seat lined. 100% imported lamb's wool. Medium gray. $9.90. [$20-25] **English-type Rib Knit Orlon Pullover.** Two front pockets. Cornflower blue. $5.97. [$15-20] **Plaid Walking Shorts.** Self fabric belt, 100% wool. Olive drab, bright blue and ivory. $6.77. [$15-20] Fall/Winter 1959.

Fall fashions for campus wear. **V-neck Pullover.** Worn with a shirt dickey. Australian wool/kid mohair/ shetland wool blend. Charcoal gray heather. $5.94. [$12-15] **Ancient Tartan Plaid Slim Skirt.** 100% imported wool, partly lined. Wine, gold and green plaid. $9.90. [$15-20] **Bulky 100% Wool Turtleneck Pullover.** Cable stitch pattern. Gold. $9.97. [$12-15] **100% Wool Flannel Walking Shorts.** Side seam pockets. Oxford medium gray. $7.74. [$12-15] Fall/Winter 1959.

Three-piece Outfit. Wool flannel weskit and skirt with cotton broadcloth blouse. Dark oxford gray. $10.94. [$35-40] **Pinwale Corduroy Jumper.** Copen blue. $5.97. **Gingham Checked Blouse.** Combed cotton, white pique collar. Blue/white/black. $2.97. [$15-20] **Hi-Rise Suspender Wool Flannel Skirt.** Tassel trim. Red. $7.97. [$40-45] **Cotton Broadcloth Blouse.** Embroidered sleeves. White with red. $3.94. [$25-30] **Two-piece 100% Worsted Wool Knit Dress and Jumper-Cardigan.** Turquoise. $16.50. [$40-45] Fall/Winter 1959.

Sporty tops and corduroy separates combine for the perfect casual outfit. The red outfit on the right is sold as a two-piece. Tops, $2.83-$3.84. [$15-20] Hi-rise capri or pedal pushers, $3.77-$3.97. [$35-45] Jumper, $5.84. [$15-20] Skirt, $5.77. [$15-20] Fall/Winter 1959.

Classic separates for fall and winter. **Block Pattern Plaid Set.** Cotton knit plaid top and matching capri pants. $8.97. [$35-40] **Classic Oxford Shirt and Wool Plaid Pants.** Shirt, $3.97. [$12-15] Pants, $6.77. [$12-15] **Schiffli Embroidery Trim Shirt.** Dacron/cotton roll sleeve. $3.97. [$12-15] **Wool/Nylon Slacks.** Self belt included. $6.77. [$12-15] Fall/Winter 1959.

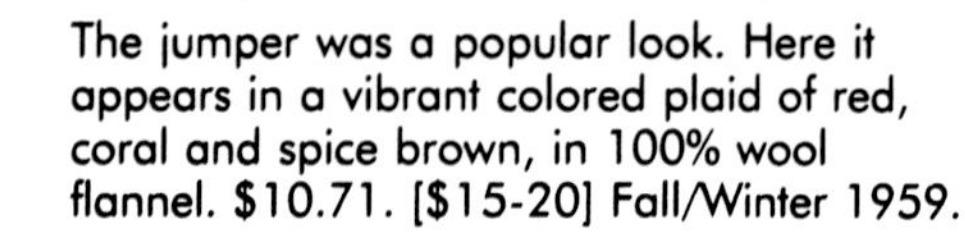

Leather Beret. Grosgrain trim, in suede or capeskin leather. Pebble beige. $3.97. [$25-30] **The Boot-About.** "The latest shoe rage for fall." Two eyelet tie. Pebble beige suede. $4.77 pair. [$20-25] Fall/Winter 1959.

The jumper was a popular look. Here it appears in a vibrant colored plaid of red, coral and spice brown, in 100% wool flannel. $10.71. [$15-20] Fall/Winter 1959.

Cotton knit mix-and-match casuals include a demi-fitted Popover top, slim skirt, striped pullover, and fully lined capri pants. Shown in teal blue. $4.84-$5.79. [$15-20] Fall/Winter 1959.

Pure wool separates include bow-tied rayon blouse, slim pants, weskit, and sheath skirt. Heather brown and peacock blue plaid. $4.77-$8.80. [$15-20] Fall/Winter 1959.

All pieces mix and match for that perfect outfit. Sweaters are 100% Hi-bulk Turbo Orlon and are completely machine washable. Shown in sweater-shirt, classic pullover and cardigan, tie pullover, and textured bulky knit popcorn stitch. $3.83-$6.84. [$20-25] Skirts and slacks are 100% wool worsted in solid bright blue and blue and antique green plaid colors. Skirts, $6.84-$8.80. [$15-20] Pants, $8.77. [$12-15] Fall/Winter 1959.

Casual Separates - Blouses

A sampling of blouses are offered on the following pages to show the many different designs and styles during late decade 1950s. Fabrics used in this time period were advertised as "carefree" due to the ease in cleaning. Many blouses were now "washable" and featured the convenience of "drip-dry." By the end of the decade, some fabric labels were beginning to read "machine washable."

Style variations are seen in collar treatments, sleeve lengths - the 3/4 length was a popular "new" style, hem treatments and trimming. Empire-line waists, hip-band "overshirt," and Italian style collars were popular. Some styles provided the versatility of sleeve-length options, convertible collars, or combination shirt/vest to be worn together or separately. One style offered the wearer the choice of wearing the blouse "as is" or completely turned around for a different style. For that perfectly matched couple, the Sears catalog also offered matching men's and women's shirts.

Left to right, top to bottom. **Casual Style Cotton Broadcloth.** Peacock blue. $3.77. [$20-25] **Pure Silk Polka-Dot.** Red dots on white. $5.97. [$20-25] **Tailored Pure Silk,** slit pocket with "arrow-head" embroidery. Stitch-outline Italian-style collar. Light gold. ¾ sleeve, $7.97. [$15-20] Short sleeve, $6.97. [$15-20] **Tucked Nylon Tricot**, sheer over opaque nylon, buttons in back. Light pink. $5.74. [$25-30] **Sizzle-Stripe Cotton Broadcloth.** Italian-style shirt. Black and white stripe. $2.83. [$25-30] **Nylon Georgette Empire-Line Blouse.** Soft gathers at midriff, back buttons. Red. $3.77. [$20-25] **Tailored Print Cotton Shirt**. Light olive and red on white. $3.77. [$12-15] **No-Iron Dacron Crepe**, tucked and ruffled bib trimmed with nylon lace. White. $5.74.[$12-15] **Striped Silk with Adjustable Rolled Sleeves.** $6.97. [$12-15] Fall/Winter 1957.

100% Acrilan Washable Jersey by Allen Blouses. *From left to right, top to bottom.* **Ivy-League Style** button down collar buttons in back also. Light/dark gray with gray trim stripe. $3.97. [$15-20] Red, $2.97. [$15-20] **Neckerchief Collar** with diagonal tucked back, gold-color pull-through holder. Turquoise blue. $3.97. [$15-20] **Sissy Blouse**, ruffled pin-tucked bib with lace. White. $3.97. [$15-20] **"String-Tie" Blouse.** Curved yoke in front and back. Light gold. $3.97. [$20-25] **Classic Simplicity.** Detachable "pussycat bow" of acetate taffeta. Black with light gray bow. $3.77. [$20-25] **Italian-style Blouse** buttons on the diagonal. Rose pink. $3.77. [$15-20] **Striped Italian-Style Blouse** "ideal choice for the girl who loves a casual way of life. Newest above-elbow sleeve length." Beige, brown and white stripes. $4.97. [$15-20] **Drawstring Blouse,** wooden toggle pin to close collar for stovepipe effect. Light amber. $3.77. [$15-20] **Back-pleated Blouse with V-yoke**, convertible peter-pan collar. Red. $4.97. [$15-20] Fall/Winter 1957.

Assorted blouses in Dacron crepe, Dacron/nylon/cotton blend, silk broadcloth, silk and acetate blend, and nylon georgette. $3.77-$6.97. [$15-20] Spring/Summer 1958.

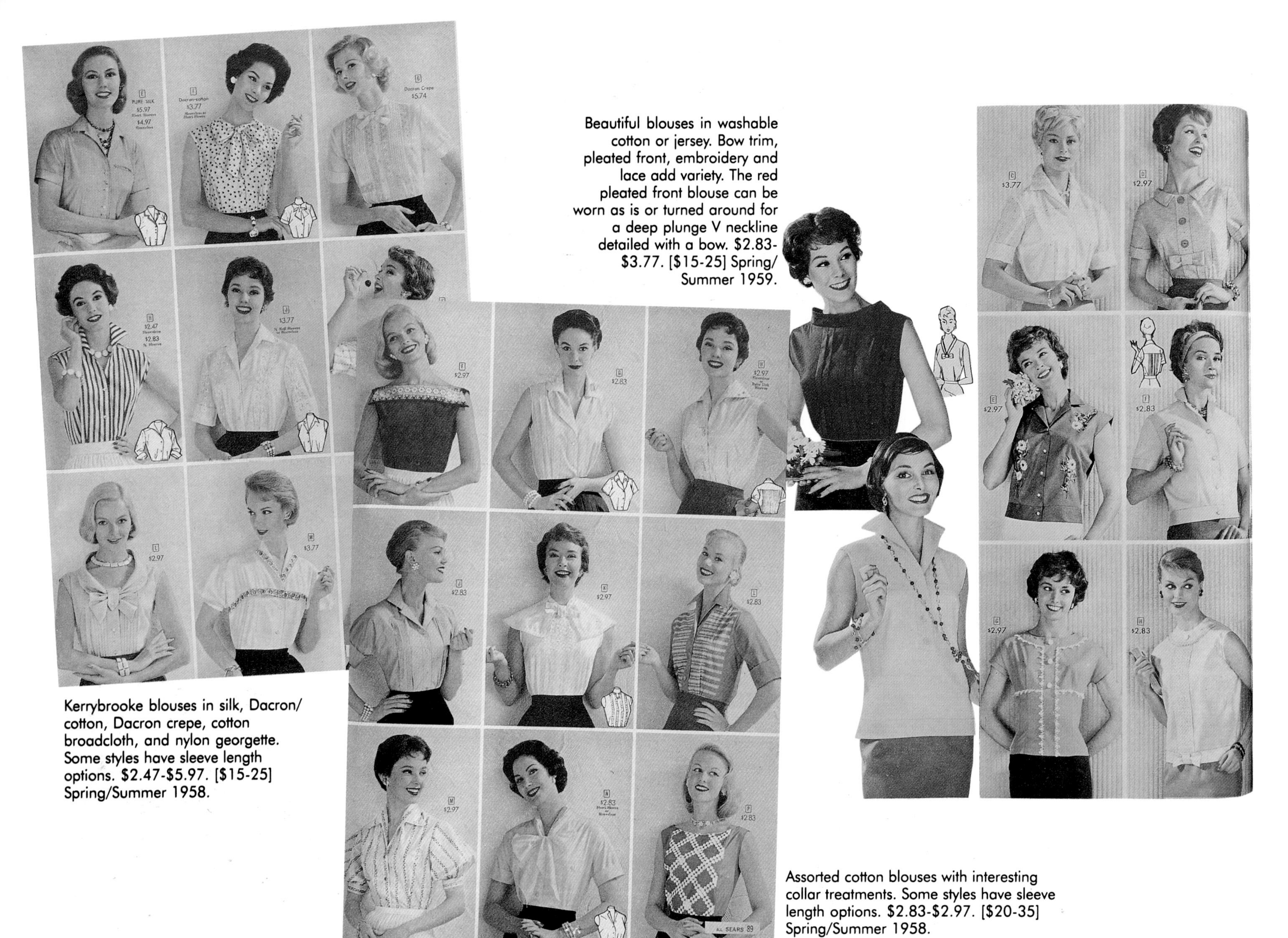

Beautiful blouses in washable cotton or jersey. Bow trim, pleated front, embroidery and lace add variety. The red pleated front blouse can be worn as is or turned around for a deep plunge V neckline detailed with a bow. $2.83-$3.77. [$15-25] Spring/Summer 1959.

Kerrybrooke blouses in silk, Dacron/cotton, Dacron crepe, cotton broadcloth, and nylon georgette. Some styles have sleeve length options. $2.47-$5.97. [$15-25] Spring/Summer 1958.

Assorted cotton blouses with interesting collar treatments. Some styles have sleeve length options. $2.83-$2.97. [$20-35] Spring/Summer 1958.

More blouse styles in drip-dry cotton. *Top left.* The "Italian-style" shirt was a popular choice. Also in style was the nautical "middy-style" designs. All in washable cotton broadcloth. $2.47-$3.97. [$20-35] Spring/Summer 1959.

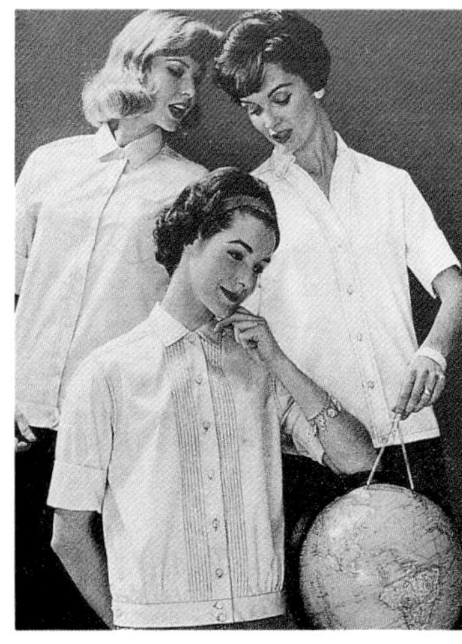

Classic shirt styles. The hip-band "overshirt" in plain and tucked front offered a comfortable but tailored fit. Vest and shirt sets were a double fashion value to be worn separately or together. In cotton broadcloth, Dacron/cotton blend, or cotton flannel. $1.86-$3.77. [$20-35] Fall/Winter 1959.

His and hers coordinated outfits for casual campus wear. **Matching Heraldic Print Shirts.** Combed Pima cotton broadcloth. Hers, short sleeves with button trimmed cuffs, $3.30. [$15-20] His is Ivy-league style, $3.90. [$25-30] Fall/Winter 1959.

Checked Shirt. Hers looks like a pullover but actually buttons all the way down. Dacron/cotton blend. Red check. Hers, $3.30. [$15-20] His, $3.90. [$20-25] Fall/Winter 1959.

Casual Separates - Sweaters

Sweater styles were as varied as blouses. Variations include collar treatments, necklines, and bow trimming. The timeless pullover and cardigan sets were very popular, as they are in the 1990s. The common theme, however, was the way they fit. In the late 1950s, sweaters were fitted to the body, and often closely fitted at the waist.

Some sweaters were "fully fashioned" meaning that the sleeve stitching is knitted directly into the sweater. A less expensive "mock fashioned" sweater has simulated stitching to make it look more expensive. For versatility some styles featured removable dickeys or collars.

Look for the similarities with today's styles.

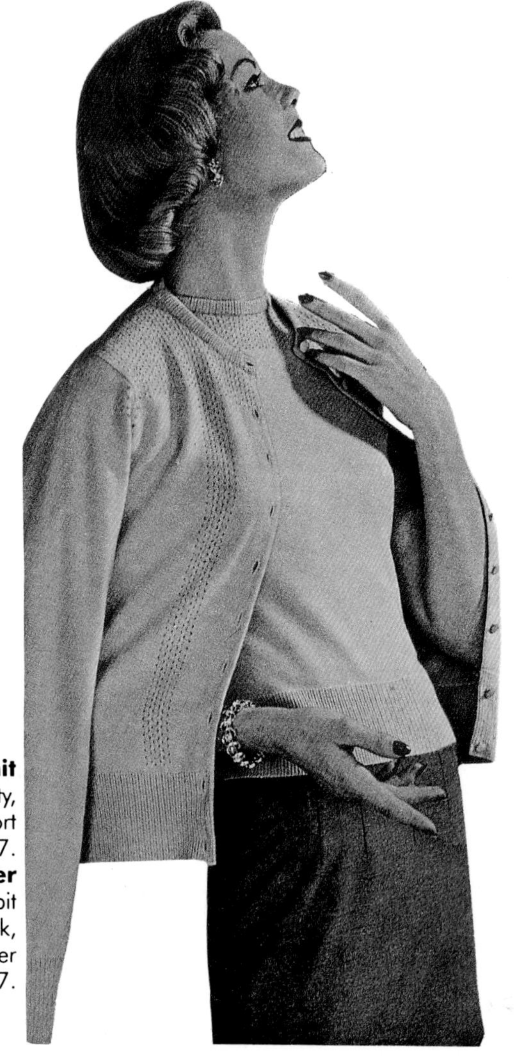

100% Hi-bulk Turbo Virgin Orlon Lacy-Knit Sweater, "the newest rage in knits. So pretty, they're collector's items..." Romantic Pink. Short sleeve pullover, $4.77. [$15-20] Cardigan, $6.77. [$20-25] Set $11.27. [$40-45] **Paneled Pullover in Part Fur Fibers.** Lamb's wool/Angora rabbit hair/nylon. Sculptured panel and mock turtleneck, ¾ sleeves. Aqua Blue. $9.77. [$20-25] Fall/Winter 1957.

Classic 100% Cashmere Sweaters. Long and short sleeve pullovers. Light blue and Geranium rose. Long sleeve, $19.67. [$40-45] Short sleeve, $15.67. [$35-40] Fall/Winter 1957.

Sears finest full-fashioned Kerrybrook Classic sweater sets.
In 100% imported cashmere.
Cardigan $18.67 [$45-50],
Short sleeve pullover, $13.67. [$35-40]
Long sleeve pullover, $15.67. [$40-45]
In part fur fiber.
Cardigan, $6.77. [$25-30]
Short sleeve pullover, $4.77. [$15-20]
Long sleeve pullover, $5.74. [$20-25]
In 100% Virgin Hi-bulk Turbo Orlon.
Cardigan, $5.67. [$25-30]
Short sleeve pullover, $3.77. [$15-20]
Long sleeve pullover, $4.77. [$25-30]
Fall/Winter 1958.

Kerrybrooke Sweaters in 100% Virgin Hi-bulk Orlon. Short Sleeve Pullover in azure blue. $2.83. [$15-25] Long Sleeve Pullover in Red. $3.77. [$15-25] Boxy Cardigan in dark green (worn over short sleeve pullover in topaz gold). $4.77. [$15-25] **Kerrybrooke Fitted Cardigan.** 100% Virgin Zephyr Wool. Medium Brown. $4.77. [$15-25] Fall/Winter 1957.

Fashion Sweaters in 100% Virgin Hi-bulk Orlon. **V-Neck Button-Front.** Braid-effect trim. Dutch blue. $6.77. [$20-25] **Shawl-collar Semi-fitted Cardigan**. Can be worn over-the-shoulder. Light beige. $5.74. [$25-30] **Dressmaker Cardigan.** Rounded collar. Rose pink. $5.74. [$25-30] Fall/Winter 1957.

The classic look pullover and cardigan never goes out of style. Available in Orlon, Ban-Lon®, or part fur fiber. Shown in spice gold and light teal. Pullover, $3.77-$4.84. [$15-20] Cardigan, $4.77-$6.84. [$20-25] Fall/Winter 1959.

Fashion Sweaters. *From left to right, top to bottom.* **¾ Sleeve Blouson.** Button-down collar converts to turtleneck. Lamb's wool/rabbit hair/nylon. Deep copen blue. $5.74. [$15-20] **Dolman-Sleeve Cardigan.** Buttons grouped in sets of 3. 100% Virgin Hi-bulk Turbo Orlon. Pink. $4.77. [$20-25] **Shawl Collar Pullover.** Lamb's wool/fur fibers/nylon. Sky blue. $3.77. [$15-20] **Raglan Sleeve Pullover.** 100% Virgin Hi-bulk Turbo Orlon. Intarsia openwork V design with matching buttons. Light cherry red. $4.77. [$20-25] **V-Neck Cardigan.** Rib-knit collar, cuffs and waistband. 100% Virgin Hi-bulk Turbo Orlon. Beige. $3.77. [$15-20] **Club-collared Slip-on.** Ribbed detail on collar and V neckline, cuffs, and waistband. Roseglow. $2.83. [$20-25] **Romantic Bedecked Cardigan.** Simulated pearls and glittering rhinestones. 100% Virgin Hi-bulk Turbo Orlon. Light blue. $7.74. [$15-20] **Scalloped Sparkle Shrug.** Simulated pearls and shimmering glass baguettes. White. $3.77. [$25-30] **Braid-trimmed Cardigan,** "beribboned, beguiling and just plain beautiful." 100% Virgin Hi-bulk Turbo Orlon. Light pink. $3.77. [$15-20] Fall/Winter 1957.

Fashion Sweaters in 100% Virgin Hi-bulk Orlon. **High-fashion Pullover.** Square neck with three-button trim. Dusty pink. $5.74. [$15-20] **Casual ¾ Sleeve Cardigan.** Lilac. $6.77. [$20-25] **Tuxedo-front Jacket in Basket-weave Stitch.** Pins detach. White. $6.77. [$15-20] Fall/Winter 1957.

Fashion Sweaters in 100% Virgin Hi-bulk Orlon. **Angora Rabbit Hair Scalloped Collar and Cuff Edge.** Deep amber with white trim. $3.77. [$45-50] **Bow Trim Bateau.** Turquoise and white. $2.83. [$40-45] **Mock Turtle Pullover.** White Angora rabbit hair trim. Coral pink. $2.83. [$35-40] **Coin Dot Sweater Set.** Contrasting intarsia dots on front and back yokes. Light gray heather. $9.97. [$45-50] **Cable-stitch Turtleneck.** Light tan. $3.77. [$15-20] **Tabbed Turtleneck Pullover.** Lamb's wool/fur fibers/nylon. Black with gray tab. $3.77. [$40-45] ¾ **Raglan-sleeve Pullover.** Mock turtle. Red. $2.97. [$15-20] **Striped Ivy-League "Shirt".** Lamb's wool/fur fibers/nylon. Gold, white and black stripes. $3.77. [$45-50] **Button and Bow Front.** White Angora rabbit hair trim, ¾ sleeves. Royal blue. $3.77. [$20-25] Fall/Winter 1957.

The term "full fashioned" as in "full fashioned sleeves" refers to the sleeve stitching knitted directly in the sweater. "Mock fashion" have simulated stitching to make them look more expensive. Shown here are assorted styles, some available in budget-minded mock fashioning which is about $2 less. Ban-Lon® or Orlon. $3.77-$8.86. [$20-25] Spring/Summer 1958.

Classic sweaters and cardigans in Ban-Lon® or Orlon in spring colors of Geranium rose, mint green, and maize. Bulky-knit sweaters were a new seasonal favorite cover-up. $3.77-$7.74. [$15-25] Spring/Summer 1959.

Paris-inspired classic set in 100% Virgin Hi-bulk Turbo Orlon. Ruby red with medium heather gray. Sweater and matching cardigan. Set, $12.97. [$25-30] Fall/Winter 1958.

Chemise Sweater. Loose-waisted, long-line cardigan with twice-bowed, rib-knit pleat panel. Lamb's wool/angora rabbit hair, nylon blend. Deep copen blue. $9.97. [$20-25] Fall/Winter 1958.

High fashion sweaters with "couterier details and textures." *From left.* **Italian-inspired Continental Look Sweater Set.** Fine gauge knit of 100% Virgin Turbo Orlon with decorative red embroidery. Heather gray. $12.97. [$40-45] **Shawl Collar Pullover.** Dickey effect Ban-Lon®. Cranberry red. $6.94. [$20-25] **Continental Style Cardigan.** Italian-influenced flat knit of 100% Virgin Turbo Orlon. ¾ length push-up sleeves, covered buttons. Teal blue. $7.97. [$20-25] Fall/Winter 1959.

His and hers fleeced cotton pullovers were perfect for fall football games. Continental collars, embroidered pocket emblem, metal buttons. White with black trim, or black with white trim. Hers, $3.90.[$35-40] His, $4.90. [$55-60] Fall/Winter 1959.

Women's sweaters in various styles. The ¾ length sleeve was a popular alternative between the full length and the short sleeve. Dickey insets or removable dickeys were fashionable and practical features. All sweaters in 100% Virgin Turbo Orlon. In teal blue, cranberry red, light beige, coral pink, cornflower blue and bright gold. $2.87-$6.84. [$25-30] Fall/Winter 1959.

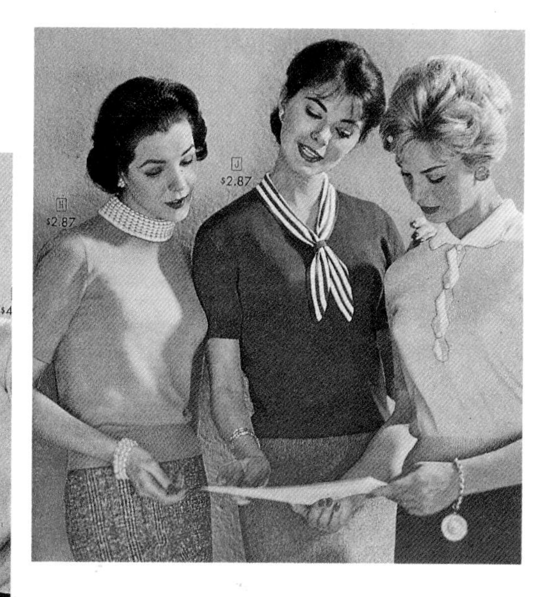

Casual Separates - Skirts and Slacks

Unlike today, skirts in the late 1950s, full or slim, were worn just below the knee. The only exception is the casual *skort* which combines a short skirt with attached elasticized pants.

Full skirts were gathered at the waist in pleated or unpressed pleated styles. Full-circle skirts were popular styles, especially for young people who enjoyed twirling around on the dance floor. Many skirts featured attached net petticoats to ensure the full flounce look.

Slim skirts, on the other hand, hugged close to the hips and tapered to below the knee. A popular choice was the wraparound slim skirt, with the added versatility of a reversible pattern. If lined, skirts were usually lined only in the seat area or from waist to hip.

Slacks and shorts were available in a variety of lengths. The most popular ones were pedal pushers, just below the knee, and the capri pants, slim and fitted with a length between the calf to just above the ankle. Denim jeans were worn with the hem rolled up. All styles accentuated the waist and hips, and followed the curve of the legs.

Wool skirts in slim and full styles. *From left.* **Gold Tweed.** Contour shaped waistband, inverted back pleat. $6.94. [$20-25] **Blue Plaid.** Unpressed pleats all around, wide simulated crushed leather belt. $8.84. [$15-20] **Houndstooth Check Slim Skirt.** Mock leather belt trimmed with skirt fabric. Black, teal blue and gold check. $7.97. [$20-25] Fall/Winter 1959.

A variety of skirts in colorful prints, all flared style with unpressed pleats, plastic patent belt included. Easy-care cotton. $3.84 each. [$15-25] Spring/Summer 1959.

Full swing sweeping circle skirts in slightly different styles. The gold skirt features rows of cording highlighting the waist and hemline. All in washable cotton. $3.97-$5.97. [$35-55] Spring/Summer 1958.

Colorful cotton skirts in geometric, floral and stripe prints. $3.84 each. [$20-35] Spring/Summer 1958.

The reversible wraparound slim skirt features a solid color on one side, embroidery-like print on the other, and is available in three lengths. $4.77. [$20-25] Women's wash and wear cotton skirts in flared styles. $2.83-$5.47. [$15-20] Spring/Summer 1959.

Pleated Skorts features all-in-one short skirt with attached elasticized pants. Shown in polka dot, solid, and striped designs. Black and white, turquoise and white, red and white. $2.83. [$30-35] Matching skirt, $3.77. [$25-30] Spring/Summer 1958.

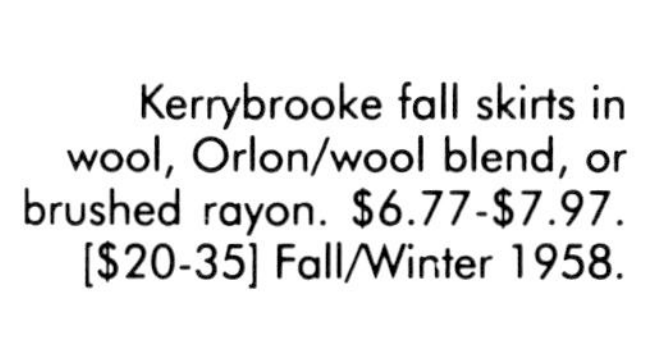

Reversible skirts in pleated and slim wraparound styles. Multicolor plaid is predominantly blue/green on one side, rust and green on the reverse. Orlon/wool blend. $8.77. [$15-20] Wraparound is red/olive plaid and solid black on reverse, both trimmed in black braid. Wool plaid, wool flannel black. $7.97. [$15-20] Fall/Winter 1959.

Reversible Skirt featured dark plaid on one side, lighter plaid on the other. Pleated Orlon/wool blend. Red plaid on white. $8.71. [$15-20] Fall/Winter 1958.

Kerrybrooke fall skirts in wool, Orlon/wool blend, or brushed rayon. $6.77-$7.97. [$20-35] Fall/Winter 1958.

Popular pants styles came in two lengths, walking shorts and pedal pushers. Shown here are two pedal pusher styles, the rolled cuff and slit-leg. All priced $3.37-$3.97. [$25-45] Spring/Summer 1958.

Capri pants, longer than pedal pushers and shorter than regular pants, were "the fun fashion of our times." An assortment of fabrics include stretchable Lastex®, cotton knit, corduroy, combed cotton sateen, and denim. Belted and unbelted styles, in stripes, prints, checks and solids. $2.83-$5.97. [$35-45] Fall/Winter 1958.

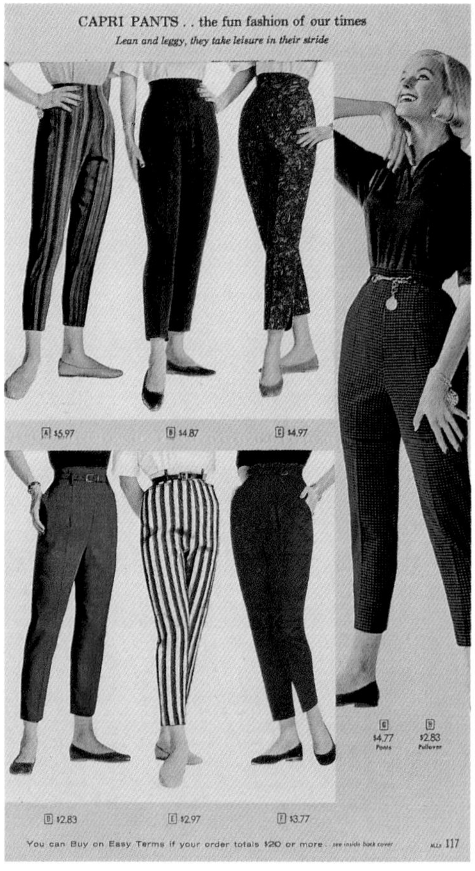

Classic roll-up jeans and pedal pushers in denim or cotton twill. The contrasting top stitching around front pockets accentuates the hips. $1.97-$2.83. [$65-85] Spring/Summer 1959.

Variations of the popular capri pants. In wool/nylon flannel, rayon/wool blend, combed cotton sheen, rayon/acetate blend, and cotton corduroy. $2.97-$4.79. [$30-45] Fall/Winter 1959.

Women's Fashions

Summer Sportswear & Beachwear

Sportswear for the resort or beach is shown here in many style variations. Nautical and boat themes were especially popular, and some styles provided similar or matching outfits for the entire family. Shorts and casual slacks for outdoor activity were available in various lengths. Skirted playsuits combined skirt and shorts, and when the skirt was buttoned, it doubled as a dress. Some mix-and-match pieces offered economical ways to create numerous fashionable outfits.

Swimwear were one-piece, long-torso, and cut low on the hip. Some styles featured adjustable torso lengths. Bathing caps and scarves offered protection from the water and wind. Hats with attached scarves offered protection from the sun.

Once again, all summer sportswear and beachwear fit closely to accentuate the shape of the body.

Mariner's Middy Pullover. Sailor collar with flag-striped bib. Cotton knit. White with navy and red trim. Women's, $2.83. [$35-40] Girls' $2.27. [$15-20] Little Girls', $1.97. [$15-20] **Best Mate Shorts.** Seaman's drop front with "officer's brass" buttons that are actually copper-plated plastic. Cotton gabardine. Navy blue. Women's, $2.83. [$30-35] Girls', $2.27. [$20-25] Little Girls', $1.97. [$15-20] Spring/Summer 1958.

From left. **Women's Island Romance Sun Set.** Cotton satin swimsuit with print cummerbund and adjustable straps. Cotton jacket in coordinating multicolor print. Sun gold swimsuit, $14.50. [$35-45] Jacket, $6.77. [$35-40] **Terry-Lined Cabana Set for Dad and Son.** Jacket in cotton striped Caribbean pattern, terry cloth lining and in collar and cuffs. Matching Swim Trunks (men's trunks are fully lined) with boxer waistband and drawstring. Sea turquoise. Man's set, $7.80. [$55-60] Boy's set, $5.74. [$35-40] Spring/Summer 1958.

Skirted Playsuit. With striped skirt on, doubles as a sundress. Detachable sash to use interchangeably. Sun yellow with gold and coral. Set, $9.97. [$60-65] **Banded Midriff Sundress.** Deep square neckline in back, bodice lined to match trim. Everglaze® cotton. Sun gold print. $7.74. [$50-55] Spring/Summer 1958.

Sun Color Corrdinates for Men. Cotton/acetate/rayon shirt with decorative woven-in vertical stripes. Wash-and-Wear pleated slacks of Dacron/rayon blend. Matching belt. Black and maize. Set, $12.50. [$55-60] *Right.* **Sunny Shirt and Shorts for Women.** Cotton print shirt and cotton poplin Jamaica shorts. Sun coral, gold and white print shirt, $3.27. [$15-20] Sun coral shorts, $2.83. [$15-20] **Lacy Straw Bag from Italy.** Woven with multicolor beads, rayon lined. Sun yellow. $5.76. [$30-35] Spring/Summer 1958.

Summertime sportswear showing different styles and lengths of pants and shorts. *Top left and clockwise.* **Classic Pedal Pushers.** $2.47. [$35-40] **Walking Shorts.** $3.77. [$25-30] **Adjustable Waist and Pants leg Pedal Pushers.** $2.83. [$35-40] **Button Trim Tabs Shorts.** $1.86. [$25-30] **Western Look Walking Shorts.** $2.83. [$35-40]. **Belted Pedal Pushers.** $2.83. [$35-40] **Ivy League Walking Shorts.** $2.83. [$15-20] **Shorty Shorts.** $1.86 .[$25-30] Spring/Summer 1958.

Assorted summertime sportswear. [$15-45] Spring/Summer 1958.

All Kerrybrooke striped pants and skirts included a striped belt. $2.83-$4.77 [$20-35]. Spring/Summer 1958.

Summer sportswear in solids and stripes. Easy-care instructions to "toss in washer, tumble or drip-dry" were becoming an industry standard. Walking shorts, pedal pushers, and capri pants were still favorites. Fabric was combed cotton Bedford Cords by Dan River. Pants and skirts have brass-buckled self belt with fleur-de-lis emblem. Rose pink, light beige, light blue and white. Tops, $1.86-$3.32. [$15-35] Shorts or pedal pushers, $3.77. [$15-45] Skirt or capri pants, $4.77. [$15-45] Spring/Summer 1959.

Casual summer wear for women. *From left.* **Coordinated Casuals.** Springmaid's combed cotton broadcloth, rick-rack and cotton braid daisy applique trim. Light peacock blue and white. Camisole or tuck-in blouse, $2.34. [$30-40] Fully lined shorts, $2.87. [$20-25] Skirt. $4.77. [$15-20] **Checked Denim Go-togethers.** In black and white, or red and white. Cotton knit blouson, $1.86. [$15-20] Cuffed shorts, $1.94. [$15-20]. Cardigan vest, $1.97. [$30-35]. Walking shorts, $2.67. [$15-20] Cotton knit shirt with denim yoke trim, $1.86. [$25-30] Pedal pushers, $2.87. [$35-45] Spring/Summer 1959.

Perfect 2-piece sets for summertime casual wear. *From left.* **Floral Print Set.** Bedford cord cotton. Light peacock blue and white. $5.64. [$30-35] **Pedal Pusher Stripe.** Blouse has removable stripe inset that matches pants. Cotton sateen. White top, black/red /white stripe. $5.74. [$30-35] **Plaid and Solid Set.** Pullover with plaid trim match walking shorts. Combed Dan River cotton. Copen blue top, copen blue/mustard gold/beige/white plaid. $5.57. [$25-30] Spring/Summer 1959.

Embroidered Duo. Blouse collar matches pedal pushers. Sheen cotton. Light emerald green. $5.84. [$45-55] **Dot and Check Trio.** Fitted overblouse, shorts and skirt worn together or separately. Percale. $6.77. [$55-65] **Striped Twosome.** Cotton knit set. Black and white. $6.77. [$35-40] Spring/Summer 1959.

Assortment of women's summer sportswear fashions. The button trim and adjustable D-rings give interesting variations to the pedal pusher. Tops, $1.86-$2.94. [$15-20] Shorts, $2.47-$2.97. [$15-20] Pedal pushers, $2.83-$2.87. [$35-45] Spring/Summer 1959.

Automatic wash and wear cotton coordinates in cornflower blue and medium beige. *From left.* **Slim Skirt and Middy Overblouse.** Skirt has side panel with two fan pleats, self fabric belt with detachable leather tab, gold-color metal buckle. Blouse has mock pocket flaps. $4.77 each. [$25-30] **Paisley Blouse.** $2.83. [$15-20] **Walking Shorts or Cuffed Pedal Pushers.** Each with self belt with detachable leather tab and gold -color metal buckle. Shorts, $4.77. [$15-20] Pedal Pushers, $5.74. [$35-40] Spring/Summer 1959.

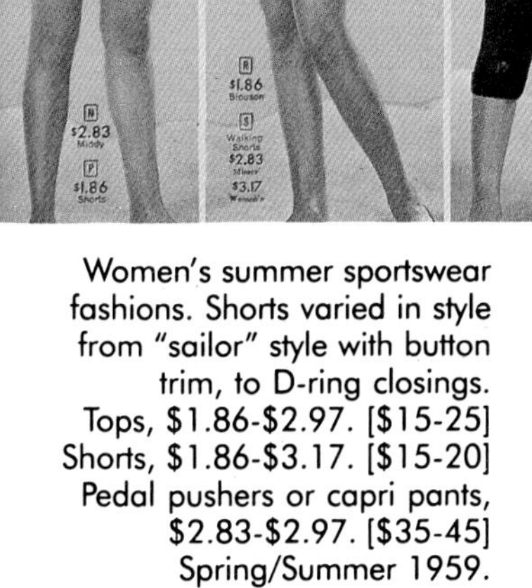

Women's summer sportswear fashions. Shorts varied in style from "sailor" style with button trim, to D-ring closings. Tops, $1.86-$2.97. [$15-25] Shorts, $1.86-$3.17. [$15-20] Pedal pushers or capri pants, $2.83-$2.97. [$35-45] Spring/Summer 1959.

Women's Swimwear. *Top, From left.* **Tapered Leg Suit with Cotton Demi-Underpant.** Elasticized faille. $14.67. [$30-35] **Shirred All-Figure Flatterer.** Elasticized faille. $14.67. [$35-45] **Star-Studded Jacquard Satin.** Criss-cross straps button at back of neck or can be tucked away. $16.86. [$65-85] *Bottom, From left.* **Knit Sheath.** Textured nylon. $10.87. [$35-45] **Celaperm Faille Check Suit.** Adjustable V-straps may be tied as cuff for sunning. $7.74. [$55-65] **Jet-Dotted Bengaline with Rayon Braid.** $8.71. [$45-50] **Halter Style Faille Suit with Cotton Terry Cloth Jacket.** Cuffed shorts. Set, $12.87. [$55-60] Suit only, $7.84. [$35-45] Spring/Summer 1958.

Kerrybrooke Swimsuits. Jacquard knit, cotton/latex, elasticized faille (acetate, cotton, rubber). Yellow nylon hood shields hair from wind and sand. Skirted styles offered a subtle cover-up and were advertised as "suitable from pool to patio." $8.76-$12.81. [$35-55] *Right.* **Exclusive Sea Star Swimsuits.** Strapless inner bra. All styles in elasticized acetate, cotton and rubber blend. Some styles had tapered legs and adjustable torso lengths. $14.74-$17.94. [$35-45] Spring/Summer 1959.

Hug-Me-Tite Swimsuit. Elasticized Bengaline (acetate/cotton/rubber blend) with permanent pleat white nylon draped cuff. Adjustable tuck away straps. Sun coral and white, $16.50. [$55-65] Spring/Summer 1958.

Women's Fashions

Maternity Wear

Sophisticated clothing for expectant mothers usually consists of a two-piece top and skirt combination, or a more casual two-piece top and pants or shorts outfit. It is interesting to note that very few skirts or slacks featured the elastic panel front that is prevalent today. To accomodate the expanding waistline, skirts had a cut-out front section (exposing the belly) and a tie front. Slacks and shorts sometimes had a multiple button adjustable waistband option, or a zip-to-fit waist, thereby allowing the piece to be worn for many months. By the end of the decade, Sears catalog began to advertise a more popular "front stretch panel."

Interchangeable tops and skirts for mothers-to-be. These maternity outfits were sold in sets of tops and coordinating skirts, and also just tops for mix and match options. In cotton and corduroy. $6.84-$8.84. [$25-30] Tops only, from $2.84-$5.84. [$15-25] Fall/Winter 1959.

Percale Top with Embossed Collar. Navy and white , or rose and white. $2.84. [$20-25] **Woven Plaid Gingham Top with White Embossed Undercollar.** Cotton broadcloth skirt. Periwinkle/blue/black/gray or Copper rust/black/gray. $3.84. [$25-30] **Satin-weave Back Acetate and Rayon.** Medium gray or Medium amethyst. $5.74. [$25-30] **Rayon and Acetate Covert Cloth.** White knitted trim. Medium gray or Light brown. $6.76. [$25-30] **Embossed Cotton, White Striped Top Tie Trim**. Navy blue/red trim, Turquoise blue, or Charcoal gray/red trim. $3.97. [$20-25] **Faille, Acetate and Rayon.** White faille collar. Navy or black. $5.94. [$20-25] **Cotton/rayon Moire Faille Top with Lamb Collar.** Skirt in ribbed acetate and rayon faille. Turquoise blue top/black skirt or Mauve pink top/navy skirt. $8.54. [$25-30] **Acetate and Rayon.** Detachable cotton lace collar dips to deep V in back.Navy or black. $8.74. [$35-40] Fall/Winter 1957.

"Our finest maternity outfit: Jacket and Skirt." Unlined jacket. Skirt with cut-out and tie-front.100% pure wool flannel. Medium amber or Red. $12.70. [$45-50] **White Rayon Broadcloth Blouse.** $4.76. Fall/Winter 1957.

Maternity casual wear. Capri pants, shorts and slacks were available with various fronts to accommodate the expanding waistline: adjustable string-tie fronts, front stretch panels, and 4-button adjustable waistbands. Tops, $2.87-$3.97. [$20-25] Shorts and pants, $2.41-$3.77. [$15-20] Sweaters. $6.97. [$15-20] Spring/Summer 1959.

Two-piece maternity dresses. Finely pleated tops were fashionable. All maternity skirts had cutout and tie fronts, leaving the enlarged belly exposed. Stretched panel fronts on skirts started to appear late decade. Sets from $5.74-$12.70. [$35-50] Spring/Summer 1958.

Shown here are four ways to accommodate the expanding belly. Walking shorts and mid-calf pants feature front stretch panels. Skirt shows the cut out tie-front. Cuffed shorts has 8-button expansion waistband and a concealed tie front. Pedal pushers has a zip-to-fit waist. $1.86-44.77. [$12-15] Spring/Summer 1958.

Elegant and sophisticated maternity wear. **Tulip Dress.** Expensively detailed petal shaped gores overlay each other, wide rounded collar gathered in back with rayon velvet streamer bow. Skirt has cut-out and tie front. In Arnel sharkskin. Pastel blue or white. $10.84. [$45-50] **Duco-print Dots.** Top has pleated back with tab-trimmed yoke. Skirt has cut-out and tie front. Shantung-textured cotton and rayon. Light blue and white, navy and white, or pink and white. $5.84. [$40-45] Spring/Summer 1959.

Women's Fashions

Loungewear & Sleepwear

Unlike everyday dresses, loungewear was advertised as "At-home fashions." These fashions included slack sets, dresses, and dusters. The difference between a duster and a robe is the length. Dusters fall somewhere just below the knee. A robe is ankle length.

Loungewear offered, most of all, comfortability of wear. These are the fashions that offered women flexibility in which to do housework. Some at-home fashions looked more presentable in case an unexpected guest showed up at the door.

Sleepwear came in many styles from cozy fleece pajamas to elegant nylon tricot peignoirs. Some styles offered matching outfits for the entire family. Others reflect popular culture with "oriental" designs and even a "chemise" style. The following pages will give an idea as to what women were wearing during late decade 1950s.

This colorful whirly skirt dress is sold as a cotton duster, to be worn at home as loungewear. Front to side tie half belt, patch pockets. Rose floral print. $6.77. [$75-85] Spring/Summer 1959.

These fashions were advertised as "stay at home" outfits for "hostess or homemaker." **Modernistic Print Outfit.** Shirtwaist blouse with roll-up sleeves, ¾ length pants, swirly wraparound skirt that ties in front. Gold and blue. Set, $7.97. [$55-65] **Dot Printed Blouson and Tapered Pants.** Blue. $5.74. [$45-50] Fall/Winter 1958.

Pinwhale Corduroy Jumper. "Considerately styled with drop seat." Red plaid. $8.71. [$45-50] **Flannelette Culotte Style.** Wide skirty pants leg. Blue. $5.74. [$40-45] Fall/Winter 1958.

Corduroy Coatchman Vest and Tapered Pants Outfit. Includes paisley print cotton blouse. Red. Set, $10.76. [$35-40] **Wrap-Style Printed Shirt and Corduroy Pants Set.** Posy-printed drip-dry cotton blouse, Knit corduroy pants, "flaring" sash. Red print, olive green. Set, $9.77. [$55-65] **Cotton Sateen Shirt and Corduroy Pants.** Open collar, roll-up sleeves, tapered pants. Aqua blue print, black. Set, $8.71. [$35-40] Fall/Winter 1958.

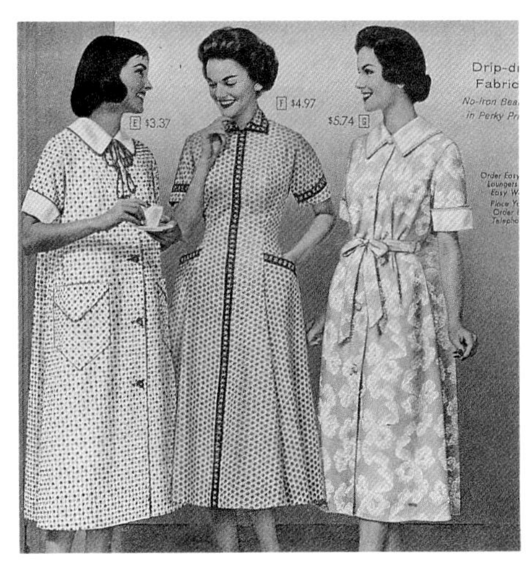

Women's Loungewear. "Redecorate for Spring with color-chic dusters". Loose and fitted styles, in multicolor prints, drip-dry, embossed cotton, and polished cotton. $2.83-$5.97. [$35-55] Spring/Summer 1958.

From left. **Navy Check Cotton Duster (short) or Red Check Robe (long).** Zip front, solid color lining, contrast piping, self belt. Duster $5.74. [$45-50] Robe $6.77. [$45-50] **Paisley Print Cotton Duster.** Corduroy collar, cuffs, pocket flaps and buttons. $8.71. [$45-50] **Celanese Acetate Duster.** Acetate lining, floral sprays accented in gold metallic and black, ¾ sleeves. Pink. $10.76. [$55-65] **Nylon Tricot Floral Print Duster.** Contrasting piping, lace trim, inverted back pleat. Aqua print on pin dot. $12.71. [$45-50] **Acetate Wrap-style Robe.** Floral applique of mock jewels, contrasting piping. Fuchsia. $8.71. [$50-55] **Nylon Chiffon Robe or Duster.** 2-button coachman style with contrasting piping and nylon lace trim. Aqua blue or Pink. Robe $12.72. [$45-50] Duster $9.77. [$45-50] Fall/Winter 1957.

Women's cotton quilted dusters and robes. *From left.* **Princess Style Paisley Print.** Acetate satin binding trim. Gold. $8.71. [$40-45] **Zippy Fitted Duster.** Check print with solid collar and lining. Navy. $5.74. [$40-45] **Coachman Long Robe.** (Also available as "Short Brunch Robe") Geometric block print. Blue. $7.74. [$45-55] Fall/Winter 1958.

Women's nylon quilted loungewear in loose and fitted styles. $7.74-$13.70. [$35-55] Fall/Winter 1958.

Women's Loungewear. Carefree cotton quilted dusters and robes to make "just staying home a fashionable affair." From $4.97-$8.71. [$50-55] Fall/Winter 1959.

Junior Loungewear. Robes for lounging "in dorm or domain." In Orlon/rayon blend, fine pinwale corduroy, quilted nylon chiffon, cotton chenille, and quilted cotton batiste. From $4.77-$9.77. [$45-50] Fall/Winter 1959.

A pleated long nylon tricot gown featuring an all around wide sweeping skirt and hand cut nylon lace was luxury nightwear. This Royal Charmode gown in ruby red or sapphire blue arrived boxed. $11.74. [$25-30] Fall/Winter 1958.

Assorted women's sleepwear in flannelette or brushed Arnel. From $2.97-$5.74. [$15-20] Fall/ Winter 1959.

Youthful monogram nightshirt, baby doll pajamas, and floral pajama set. In cotton broadcloth, cotton/rayon blend, and brushed Arnel. $3.77-$5.74. [$20-25] Fall/Winter 1959.

Daisy Muu Muu. Cotton. $2.83. [$15-20] **Granny Gown.** Cotton batiste floral with nylon satin nylon streamers. $3.77. [$15-20] **Duster and Waltz Gown Set.** Wear duster three ways: loose, half-belted, sash all around. Set, $7.74. [$25-30] Spring/Summer 1959.

For the pajama party, a monogram tailored jamarette set, pinwale corduroy duster, and a mandarin style pajama set. All cotton. $4.77-$6.77. [$25-35] Fall/Winter 1959.

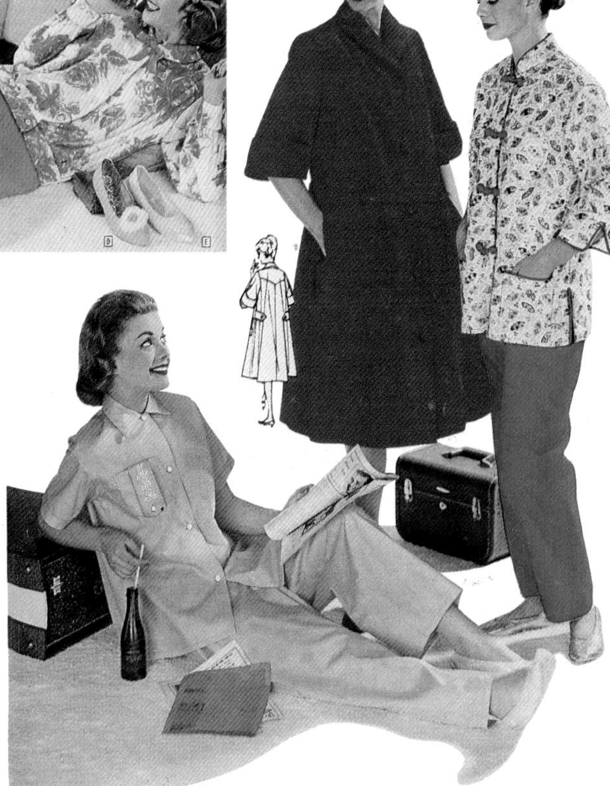

Regal blue cotton flannel loungewear for the entire family – "soft, warm loungers that spell majestic comfort from the littlest 'prince' to the 'king' himself." A warm 1950s family feeling! $2.37-$3.77. [$20-35] Fall/Winter 1959.

Mother and daughter pajama loungers in brushed knit blend of Celanese Arnel and rayon.$2.83 [$15-20] for littlest girls to $4.77 [$20-25] for women. Matching pompon slippers from $1.36-$1.66. [$20-25] Fall/Winter 1958.

Border Print Duster & Pajama. Floral design flannelette. $3.77. [$15-20] **Japanese Inspired Robe & Pajama.** Flannelette wrap-around robe. Mandarin-collar pajama. $3.77. [$15-20] **Embroidered Oriental Motif Robe & Pajama.** Flannelette wrap robe, silver-white nylon satin piping, mandarin-collar pajama. Robe $4.77. [$25-30] Pajama $3.77. [$20-25] **Printed Plisse Challis Duster & Butcher Boy Pajama.** Cotton/cotton lace, color piping trim. $4.77. [$20-25] **Red & White Print Short Lounge Coat & Ski Cuff Pajama.** Flannelette Peter Pan collars, wide sleeve cuffs. $2.83. [$15-20] Ski Cuff Pajama Trousers has wide elastic boxer waistband. $3.77. [$15-20] Fall/Winter 1957.

Adding to the family fun, "court jesters" loungewear for mother-daughter "fun-filled dreams." Sanforized flannelette. $2.67-$3.77. [$20-25] Fall/Winter 1959.

Women's Sleepwear. *Top left and clockwise.* **V-neck Combed Cotton Knit.** Batwing sleeves, Peacock blue. $4.77. [$15-20] **Turtle-neck Blazer Stripe.** Combed cotton knit, Red and white. $3.77. [$15-20] **Turtle-neck Ski-Cuff with Floral Embroidery.** Flannelette, Maize. $3.77. [$15-20] **Border Print Ski Cuff.** Flannelette, printed square dots on white with multicolor border. $3.77. [$20-25] **Coat-front Print.** Geometric pattern and pin dot, edged with piping and lace. Flannelette, Light blue and white. $2.83. [$15-20] **Blazer Stripe Print.** Ivy league collar. Flannelette, Rose and gray on white. $2.83. [$25-30] **Shirt-tail Coat Flannelette.** Blue. $3.77. [$15-20] **Man-tailored Flannelette with Tie Belt. Pink with white piping.** $3.51. [$15-20] Fall/Winter 1957.

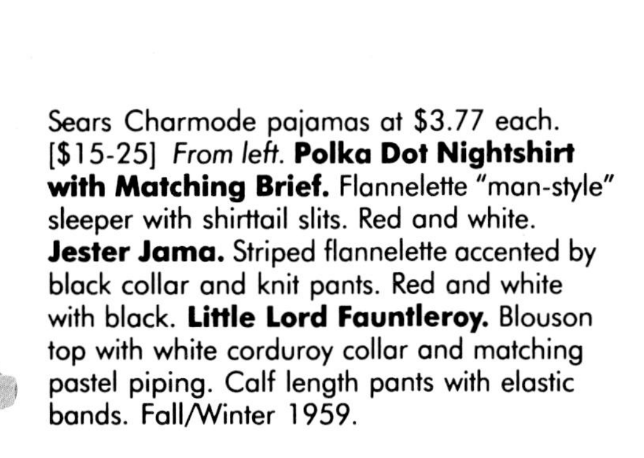

Sears Charmode pajamas at $3.77 each. [$15-25] *From left.* **Polka Dot Nightshirt with Matching Brief.** Flannelette "man-style" sleeper with shirttail slits. Red and white. **Jester Jama.** Striped flannelette accented by black collar and knit pants. Red and white with black. **Little Lord Fauntleroy.** Blouson top with white corduroy collar and matching pastel piping. Calf length pants with elastic bands. Fall/Winter 1959.

It's hard to believe that women used to iron sleepwear. These styles were advertised as no-iron cotton and wash 'n wear batiste. Short and long styles. $3.77-$8.71. [$15-20] Spring/Summer 1958.

Assorted ladies' nightwear in sanforized flannelette. *Left to right.* **Leopard Striped Outfit**, $7.75. [$30-35] **Oriental Style Set**, Pajama, $3.77. [$20-25] Robe, $4.77. [$25-30] **"Collegiate Cardigan" Set**, $5.74. [$15-20] In cotton suede, **Block Print Duster and Matching Pajama**. $3.77 each. [$15-20] Fall/Winter 1958.

Sporty Ski-cuff Pajama. Peacock blue. $3.77. [$30-35] **Ship Ahoy Set.** Choice of red or blue trousers. $4.77. [$15-20] **Tailored Printed Top Set.** Red, navy, and gray top, gray trousers. $3.77. [$15-20] **Striped Ski Pajama.** Red. $3.77. [$20-25] Fall/Winter 1958.

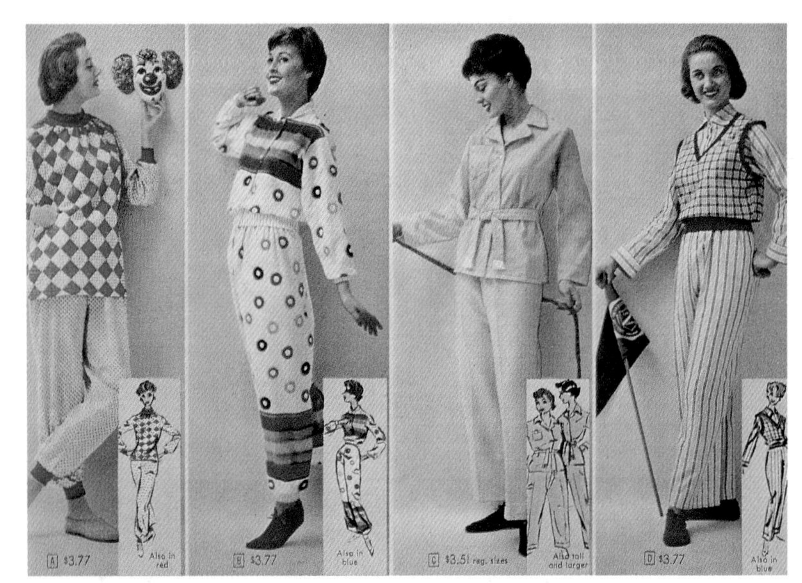

Sleepwear reflected popular circus themes, as well as the "man-tailored" look and the collegiate "sweater girl" look. All in sanforized flannelette. $3.51-$3.77. [$20-35] Fall/Winter 1958.

Sleepwear featured the always popular floral and stripe themes. A chemise-look style followed the fashion trend of the year. All in sanforized flannelette. $2.73-$3.77. [$20-25] Fall/Winter 1958.

Women's Fashions

Lingerie

A woman is never quite dressed without adequate foundation wear. In this section, a sampling of women's undergarments are shown.

In the late 1950s, women did not go after the "natural" look. Bras accentuated and brought the breasts to a sharp point. Rotation bras, with stitching in concentric circles, served this purpose. The catalog states "wonder-working circular and sunburst stitching molds and accentuates the bustline in flattering contours." It is interesting to note that young girls' bras were advertised to be "just like mother's." How many teens today are concerned about having a matching bra with their mothers?

Garters attached to briefs or girdles were common because they were needed to hold up nylon stockings. Panties and briefs were cut low on the hips but came in a variety of fabrics and styles, including those sold in sets as days-of-the week briefs. Girdles sometimes had extra panels attached to the front to smooth out the dress lines.

Slips came in a variety of styles. The long straight slip was perfect for under the "new" chemise look, but layers of net were a must to "pouf" up full swing skirts. Some styles boasted a full 180-inch sweep, or three full layers of petticoat.

Take a look at the undergarments that completed every woman's wardrobe.

Full whirly skirts required full sweep petticoats. "Ideal for square dancing or just walking with a swish in your step. " White drip-dry cotton with eyelet embroidered batiste. $2.83. [$35-40] Spring/Summer 1958.

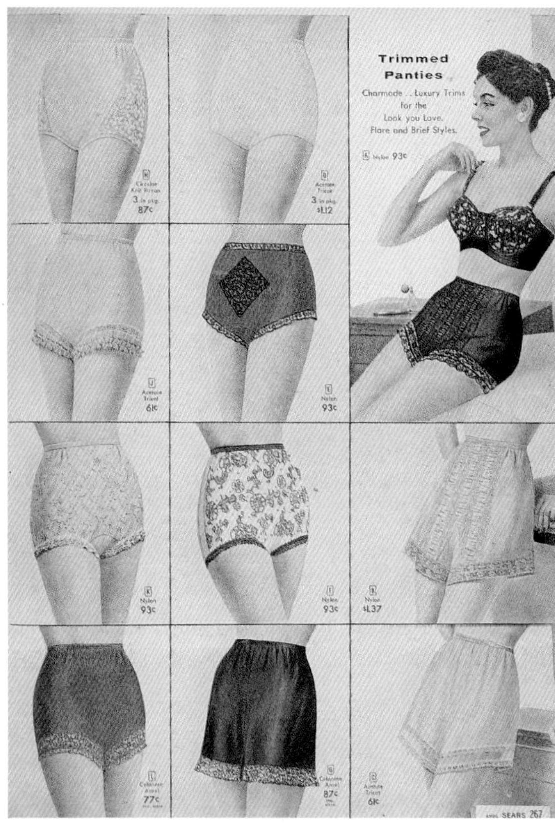

Assorted Fancy Flare Leg or Brief Styles. Knit rayon, acetate tricot, nylon, celanese arnel, assorted lace trim. $.61 to $1.37. [$5-15] Fall/Winter 1957.

Women's panties were available only in "brief " style, cut low on the legs. Shown are boxed sets of underpants in various fabrics and, of course, in popular "days of the week" design. Panties sold 6 in a box, except for "days of the week" which are packaged in box of 7. $1.97-$4.77 box. [$25-30] Fall/Winter 1959.

Assorted Tailored Panties. *From left.* **Days of the Week Briefs.** In three run-proof fabrics rated good, better, and "our best." Set of seven, from $2.37-$4.77. [$25-30] **Coordinated Chemise and Panties.** Satiny-stripe rayon tricot. $.73-$1.47. [$5-12] **Garter and Elastic Leg Panties.** Cotton mesh, acetate tricot. $.77-$1.17. **Run-A-Bout Briefs.** Combed cotton. Each $.44 or 3 for $1.29. Fall/Winter 1957.

Women's Girdles. Note the extra paneled zip front. From $6.97-$12.72. Fall/Winter 1957.

Assorted Bras. Note the matching mother-daughter "rotation" bras. "Wonder-working circular and sunburst stitching molds and accentuates the bustline in flattering contours." From $1.40-$3.77. [$15-25] Fall/Winter 1957.

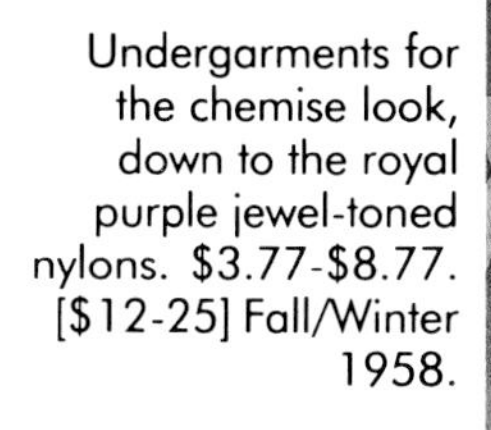

Undergarments for the chemise look, down to the royal purple jewel-toned nylons. $3.77-$8.77. [$12-25] Fall/Winter 1958.

Nylon Tricot Trio with Pleats and Lace. Slip, $8.71 [$12-15]. Petticoat, $5.74. Flare Leg Panty. $1.86. [$8-12] Fall/Winter 1957.

Nylon Tricot Trio with Pleats and Scalloped Lace. Elastic Leg Brief, $1.86. [$8-12] Petticoat, $4.77. [$10-12] Slip, $7.74. [$12-15] Fall/Winter 1957.

Permanently Pleated Sheer Nylon Tricot with Nylon Lace. Petticoat in Graduated Tiers, $12.69 [$15-20]. Continental Brief, $2.34. [$10-12] Fall/Winter 1957.

Nylon and lace luxury lingerie. From flare leg panty for $1.86 [$10-12] to a coordinated peignoir set, $16.65. [$45-50] Spring/Summer 1958.

Sears Fullest Bouffant Petticoat with a 75-yard sweep of nylon point d'esprit net ruffled in four tiers. Available in three lengths, short, regular, and tall. $5.74. [$30-35] **Cotton Whirling Petticoat** with a 180-inch sweep, eyelet embroidered ruffle. $2.83. [$15-20] Fall/Winter 1958.

Women's Lingerie. Pretty party bouffant petticoats in beautiful colors to "prop out your skirts." In nylon net with nylon tricot. $2.97-$5.97. [$35-45] Fall/Winter 1959.

Women's Fashions

Fashion Accessories

An outfit, or "costume" as the word was used in the late 1950s, is not complete without accessories. Hats, gloves, scarves, bags, shoes, cover-up shawls, and even a matching umbrella all serve to put together that total look.

The well-dressed woman often wore matched accessories. It was not unusual for a woman to wear matching polka dot hat, gloves, shoes, and scarf, plus carry a polka-dot handbag. Removable collars, cuffs, and dickeys added variety to a dress. Nylon stockings had seams down the back of the leg, with dark heels and soles. Although seamless stockings were beginning to appear in the late 1950s, most stockings sold were the seamed variety.

A hat was essential when going out, no matter what the occasion. Each woman's wardrobe should contain several styles of hats. The assortment of headcoverings on the following pages will astound you. Each season's catalog featured page after page of hats, each adorned with rayon netting, feathers, fabric flowers, or a decorative jeweled ornament. The hat pages were not printed in color, perhaps due to the fact that they were made in relatively neutral shades of blue, brown, beige, black, or red. After all, a hat should match several different outfits!

Fashion accessories are essential for a complete look. Here is a sampling of what women wore during late decade 1950s.

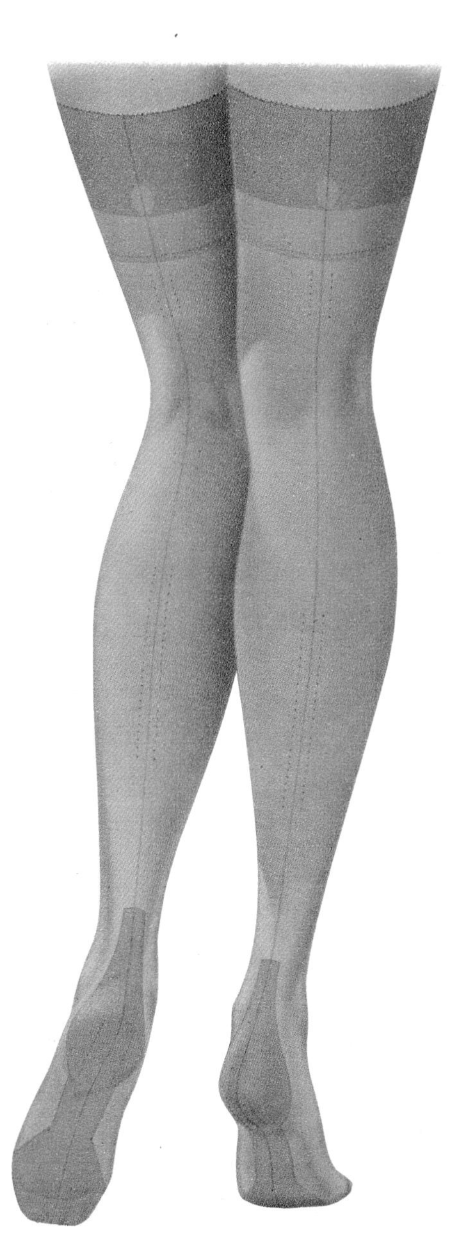

Neckwear add-ons included collars, dickeys and scarves. Some have matching cuffs. $1.27-$3.47. [$20-35] Spring/Summer 1958.

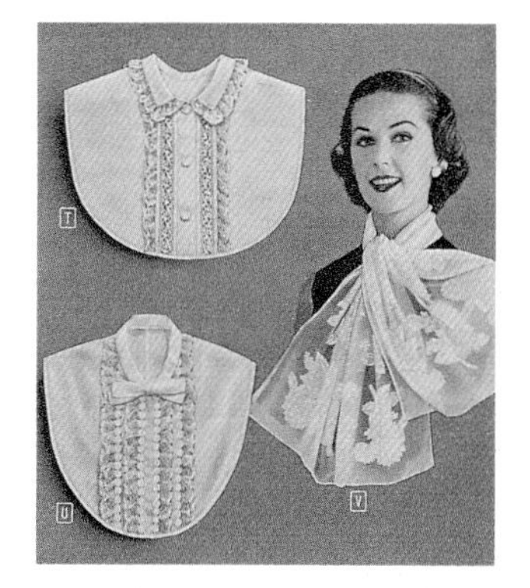

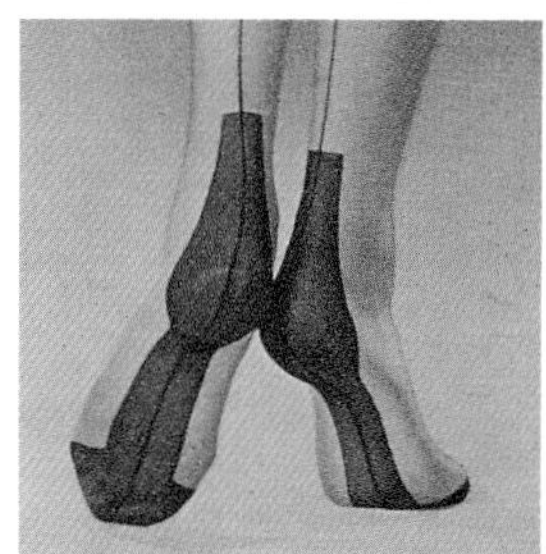

These nylons featured cotton soles and stretch cotton top. Seamed stockings with dark soles and heels were the norm, although seamless and nude heels were available. All worn with garter belt. From $.50 - $1.37. [$5-12] Spring/Summer 1958.

Popular polka dots appeared in shoes, bags, and many other fashion accessories. $.97-$7.47. [$25-45] Spring/Summer 1958.

Boleros, stoles, capes, and shawls were elegant ways to dress up an ordinary outfit. These accessories were made from crush-resistant rayon with the feel of velvet, as well as satin and lace. Priced from $1.97-$6.47. [$45-50] Fall/Winter 1959.

Flower-themed accessories were popular for spring. From a belt for $1.47 [$25-30] to a hand painted stole for $4.77. [$45-50] Spring/Summer 1958.

Women's Fashion Accessories. *Center.* **New Fur-Look Cowl Cape.** Orlon/Dynel, acetate satin lined, front self button ornament conceals crochet hook/eye clasp. Pearl white, platinum gray, or fawn (light brown). $15.97. [$25-30] *Top left and clockwise.* All styles in black. **Rayon Velvet Shrug with Studded Rhinestones.** Acetate taffeta lined. $5.97. [$45-50] **Wing-Sleeve Bolero.** Rayon velvet collar, acetate taffeta lined. $6.97. [$45-50] **Rayon Velvet Blouson Cardigan.** Push up wing sleeves, glittering buttons. $9.47. [$55-65] **Sheer Lace Bolero.** Acetate/nylon. $1.97. **Imported Lace Stole from France.** Silk net with rayon embroidery, scalloped edges. $2.97. [$35-40] **Rayon Velvet Cuddle Cap.** Rhinestone decorations at ears, rayon/acetate faille lined. Black. $1.97. **Mink Collar or Head Clip.** Dark ranch mink tails. Collar, $9.97. [$45-50] Head Clip, $5.97. [$20-25] **Rayon Velvet Stole.** Studded with rhinestones, acetate taffeta lined. $5.97 [$30-35]. Fall/Winter 1958.

Chemise tie or scarf to wear many ways. In double nylon tricot with a sliding bow. White, ruby red, autumn tan, or sapphire blue. $1.97 [$15-20]. Fall/Winter 1958.

Women's Dress Shoes. Assorted styles, $6.97-$7.74. [$35-45] Fall/Winter 1957.

Classic styles for ladies in black calf, kidskin, or suede. *Top left and clockwise.* **Slip-on Bowed Pump.** $7.74. [$35-40] **Dressy Sandal with Instep Strap.** $7.74. [$45-55] **Gypsy Tie Shoe.** Kidskin with patent leather. $7.74. [$35-40] **Simple Sweater Pump.** $6.79. [$45-65] Fall/Winter 1958.

Assorted summer shoes, some with matching bags. The wedge heel was a favorite for casual wear. $3.77-$5.74. [$35-65] Spring/Summer 1958.

Cut-out tiny openings were decorative touches to the classic pump and "this year's fashion hit, the T-strap!" In black calf or suede. Pump, $6.97 pair. T-strap, $7.74 pair. [$55-60] Fall/Winter 1958.

Women's Hats. *Top left.* **Fur-Effect Rayon Plush.** Rhinestone ornament. $5.74. [$45-50] *Center.* **Wide Oval Brim Beaver Fur Felt.** Rayon satin folds around crown, handmade motifs of gold bullion. $9.97. [$55-60] *Top right.* **Rabbit Fur Full Blouse Crown.** Rayon satin ripple brim, red rose ornament. $7.97. [$35-40] *Lower left.* **Beaver At A Peak.** Pheasant feather ornament. $7.97. [$35-40] *Lower right.* **Glittering Beaver Fur Pillbox.** Nylon veiling with metallic flecks. $5.97. [$20-25] Fall/Winter 1957.

For that perfect hairdo before donning the hat. Spring/Summer 1958.

Women's Hats. *Top left.* **Rayon Velvet Side Sweep.** Brim turned back and draped at side. Ornamental whips with rhinestone nuggets. $5.74. [$35-40] *Center.* **Cascade of Feathers.** Deep curved dome of Panne rayon velvet. $6.97. [$35-40] *Top right.* **Silky Beaver Fur Felt.** Soft bumper shape, rose with rhinestone dew-drops. $5.97. [$45-50] *Lower left.* **Rayon Velvet Ripple Brim Cloche.** Braided band of velvet and rayon satin. $5.97. [$25-30] *Lower right.* **The High Fez.** Fine fur felt, two pheasant feathers caught with rhinestone ring. $5.97. [$45-50] Fall/Winter 1957.

Women's Hats. *From left to right, top to bottom.* **Blouse Crown Beaver Fur Felt.** Rhinestones on grosgrain band. $5.97. [$35-40] **Silky Beaver Fur Felt Dome.** Long pheasant quill. $5.97. [$25-30] **Beaver Fur Felt High Crown Pillbox.** Large rhinesone snowflake ornament. $5.97. [$25-30] **Bare-The-Brow Wool Felt Pillbox.** Rayon satin bows. $3.97. [$20-25] **Swagger Fedora.** Wool felt with wide 2-tone grosgrain band. $3.97. [$25-30] **Bumper Pillbox.** Fluff-brushed wool felt, rayon satin pleated bow, rayon veil and rhinestones. $3.97. [$30-35] **Notched Cut-Out Wool Felt Cloche.** Rayon satin cuff band, rayon veil. $3.77. [$20-25] **Ombre Wool Felt Cloche.** Draped with shaded rayon jersey. $4.97. [$35-40] **Brushed Wool Felt Deep Cloche.** Wide band of rayon braid and glitter pinwheels. $3.77. [$20-25] **Fluffy Marabou High Crown Cloche.** Wool felt with soft cushion brim. $3.97. [$20-25] **Draped Brushed Wool Turban.** Beaver-like texture, pheasant quills and rhinestones. $3.97. [$35-40] Fall/Winter 1957.

Women's Hats. *From left to right, top to bottom.* **Feather Profile Shell.** Rhinestone touches. $3.97. [$40-45] **Wide Brim Circle of Marabou.** Rayon velvet. $4.97. [$35-40] **Side Swept Pompon.** Rayon velvet with rayon satin band, big ostrich pouf. $5.97. [$40-45] **Side Draped Profile Cap.** Rayon Velvet with rhinestone V ornament, rayon veil. $3.77. [$25-30] **Rhinestone Starburst Soft Pleated Hat.** Velvet and rayon satin. $3.97. [$25-30] **Velvet Side Pouf.** Softly draped crown. $3.77. [$40-45] **Big-Bow Beret.** Rayon velvet with self bow. $3.77. [$30-35] **Shimmer of Sequins Cap.** Rayon velvet with sequin scrolls. $3.97. [$20-25] **Twisted Turban.** Rayon velvet with rayon satin and rhinestone ornament. $3.97. [$20-25] **Curly Feather Shell.** Rhinestones touches. $3.97. [$20-25] **Three-Tier Pillbox.** Rayon velvet with rayon satin, rhinestone ornaments. $3.77. [$20-25] Fall/ Winter 1957.

Women's Head Clips. *Top left and clockwise.* **Rayon Velvet Feather Clip.** Cluster of feathers on one side. $2.37. [$25-30] **Profile Clip with Sequins.** Rayon velvet bands with cluster of matching sequin ornaments on side. $2.37. [$20-25] **Eye-Shadowing Clip.** Wide rayon velvet band with fine mesh rayon veil sprinkled with rhinestones. $1.86. [$15-20] **Feather Clip.** Worn "either across your brow or circling your chignon." Rayon velvet band with feathers. $2.37. [$20-25]Fall/Winter 1957.

From top. **Fur-Look Pillbox.** Rayon plush with rhinestone ornament. $3.97. [$15-20] **Fluffy White Rabbit Fur Half-Hat.** "Red rose in the snow." $2.97. [$20-25] Collar and Hat. $3.97. **White Rabbit Fur Cuddle Cap.** Rayon satin ties. $2.97. [$20-25] Collar and Cap. $3.97. [$15-20] **Fur-Look Fabric Cloche.** Rayon pile pressed to look like broadtail fur. Deep scoop shape with rhinestone ornaments. $2.97. [$15-20] Fall/Winter 1957.

Women's Scarf Hats. *Left.* **Ascot Hood.** 100% Orlon, plaid on one side, solid on other. $2.27. [$20-25] **Plaid Stole Cap.** Wool/rayon with wool fringe. $2.27. [$20-25] **Scarfed Visor Cap.** All wool jersey. $1.86. [$20-25]

Knitted Caps. *Top right and clockwise.* **Hand Crochet Wool Toque.** Bulky knit with diagonal ribbing, three rhinestone ornaments. $3.77. [$15-20] **Popcorn Stitch Knit Wool Cuddle Cap.** Dyed mouton-processed lamb chignon. $2.47. [$15-20] **All Wool Helmet.** Hand crocheted ruffle top with simulated pearls. $2.97. [$15-20] **Angora and Wool Knit Cap.** Wool/Angora rabbit hair, chignon effect in back, plastic flower ornaments. $2.27. [$15-20] Fall/Winter 1957.

From left, top row. **Dome Pillbox.** Wool jersey with attached scarf. $2.27. [$15-20] **Shirred Jersey Turban.** All wool, high draped backward slant style. $2.97. [$25-30] **Hand Crocheted Beaded Wool Cap.** Curved up in back to a loop fringe crown. $2.97. [$15-20] **Open Crown Cuddle Cap.** 100% wool, stitched with white Angora rabbit hair, simulated pearls. $1.97. [$20-25] *Row 2.* **Bead Trimmed Cuddle Cap.** Simulated pearls. Wool jersey or cotton velveteen. $1.86. [$20-25] **Pony Tail Helmet.** Simulated pearls, bandeau clip. Wool jersey or cotton velveteen. $1.86. [$25-30] **Tweed-Effect Turban.** Wool/rayon/cotton knit on wool jersey crown. $1.97. [$20-25] **Fluffy Toque with Beehive Shape.** Circle top, knit of Angora rabbit hair/ wool reinforced with nylon. $2.27. [$15-20] Fall/Winter 1957.

Top left and clockwise. **Profile Toque with Feather.** Shaggy brushed wool felt pierced with pheasant quill. $3.47. [$15-20] **Off The Face Brim.** Soft wool felt with stitching, rayon grosgrain band and loops. $2.83. [$25-30] **Swathed 'round with Veiling Wool Felt Cloche.** Rayon mesh veiling flecked with metallic sparkles. $2.97. [$15-20] **Bow Front Bumper Pillbox.** Brushed wool felt, flat tailored bow in rayon satin to match felt. $2.83. [$15-20] **Scalloped-edged Cloche.** Wool felt, crown woven through with strands of white Angora rabbit hair yarn, finished with tassel. $2.83. [$15-20] **Bows in a Row Brushed Wool Felt Pillbox.** Three crisp matching bows of rayon slipper satin. $2.97. [$15-20] Fall/Winter 1957.

From left. **Circle of Fluffy Maribou.** Rayon panne velvet. $3.47. [$30-35] **Rayon Slipper Satin Shell.** Decorated with simulated pearls. $2.83. [$15-20] **Large Rippled Brim.** Rayon velvet draped brim, two big rhinestone ornaments. $3.47. [$20-25] **Rayon Velvet Turban Pillbox.** Banded with self-trim, rhinestone ornament, rayon veil. $3.47. [$30-35] Fall/Winter 1957.

From left, top row. **Wings in Flight Rayon Velvet.** Matching feathers, veil. $2.97. [$25-30] **Draped Rayon Velvet Mushroom.** Rhinestone ornaments. $2.97. [$15-20] **Pointed Petals.** Rayon velvet with rayon veil. $2.83. [$15-20] **Curved Sequin Pillbox.** Rayon veil that ties back. $2.97. [$15-20] **Ostrich Pompon Toque.** Rayon velvet. $2.97. [$25-30] **Buckled Turban Pillbox.** Rayon velvet rayon satin folds, rhinestone buckle, rayon veil. $1.86. [$25-30] **Level-headed Pillbox.** Rhinestone trim, rayon veil. $2.83. [$15-20] **Rayon Panne Velvet Cap.** Rayon satin tubing forms side clips, rayon veil, feather. $2.83. [$25-30] **Swirling Sequins Calot.** Rayon velvet. $2.83. [$20-25] **Wide Brimmed Rayon Velvet.** Rayon veil ties back. $2.83. [$15-20] **Profile Brimmed Rayon Velvet Hat.** Bow at side, feather spray top. $2.83. [$20-25] **Ribbon-Trimmed Cap.** Rayon velvet with wide rayon grosgrain pleated ribbon. $2.83. [$20-25[Fall/Winter 1957.

Women's suit jackets varied in style and length. *From left.* **Wool/nylon Target Cloth Suit.** Polka dot pattern acetate taffeta on lining and ascot. Jacket about 22 inches long. Red with red and white pattern lining. $12.70. [$65-70] **Pure Wool Flannel with Fur Collar.** Rhinestone-centered buttons at fitted jacket waist. Jacket about 20 inches long. Copen blue with blue fur. $22.70. [$65-70] **Iridescent All Wool Tweed.** Puritan collar, acetate taffeta lined. Jacket about 21 inches long. Iridescent gold. $16.90. [$65-75] **Flared Suit.** Wool target weave, flared skirt. Jacket about 19 inches long. Coral red. $19.70. [$85-90] **Oval-shaped Soft Plaid.** Pure wool, acetate satin lined. Jacket about 34 inches long. $29.70. [$55-60] Spring/Summer 1959.

Fall suits reflecting a "french influence." *From left.* **Two-piece Walking Suit.** Oval shaped wool/nylon plaid brief coat, rayon/acetate taffeta lined, wool flannel chemise dress. Copper rust and gray. $29.67. [$65-75] **Three-piece Costume Suit.** Bold plaid wool jacket, acetate taffeta lined. Wool flannel skirt and wool knit jersey blouse. Sapphire blue and black. $19.67. [$65-75] **Two-piece Chemise-look Suit.** All wool flannel, inverted center pleat and tucks in black, flanged band all around jacket. Bright red. $17.67. [$75-95] **Oval Shaped Walking Suit.** Wool/nylon multicolor fleck tweed, shoulder-wide yoke brief coat, and matching skirt. Wine, red and black tweed. $24.67. [$75-95] Fall/Winter 1958.

The fur trimmed suit was a new look for fall. *Top, left and right.* **Black-dyed Red Fox Collar.** (Fur origin: Canada). Wool plaid broadcloth, rayon/acetate crepe lined. Sapphire blue and black. $39.67. [$75-85] **Glamorous Dyed Mink Sides Collar.** Wool craw-foot tweed, jacket lined in acetate taffeta. Brown and white. $27.67. [$65-75] *Bottom, left and right.* **Spotted Ocelot Fur-effect Fabric *(rayon/cotton).*** Lining and collar on wool/alpaca suit. $27.67. [$95-115] **Dyed Mouton Processed Lamb Shawl Collar.** Wool flannel suit, lined in acetate taffeta. Bright red, black fur. $22.67. [$95-110] Fall/Winter 1958.

Women's Fashions

Outerwear - Jackets & Coats

Styles in outerwear include jackets and coats. Some styles remain fashionable today and are sought after by collectors of vintage clothing. Though coats sold today vary little from those in the late 1950s, there are some differences.

The material with which a coat is made is one major difference. It was not unusual to use actual animal fur in the coat or in a fur-trimmed collar or cuffs. In fact, fur was so desirable that a new fur-like fabric called *Wink* became a popular and less-expensive choice. *Wink* was a man-made fabric with the look and feel of mink and was manufactured as full-length coats, shorter coats, and stoles. Cloth coats with fleece-pile lining also were favorites, mainly for their warmth.

Another difference is in the way the coats were named. How many of today's generation know the term "car coat?" A car coat was slightly longer than a jacket but shorter than a full-length coat, just perfect for getting in and out of a car. "Toppers" were shorter coats, and "Greatcoats" were full length cover-all coats.

The following outerwear are grouped by style and arranged by year to give an appreciation of seasonal styles and color trends.

Hooded Capeskin Blouson Jacket. Draw-string blouson, Orlon fleece lined jacket and hood. Red with white fleece lining. $32.50. [$65-70] **Topper-length Capeskin Jacket.** Orlon fleece collar and lining. Periwinkle blue with white fleece lining. $41.00. [$45-55] **Topper-length Suede Leather Jacket.** Treated to resist water-spotting, rayon twill lined. Beige Suede. *Also in Capeskin.* Beige or white. $29.97. [$45-55] **Classic Suede Leather Jacket.** Fully lined with rayon twill. Copper Rust Suede. *Also in Capeskin.* Beige, white or light blue. $23.67. [$45-55] Fall/Winter 1957.

Popular Style Car Coat in 100% Wool Tweed. Alpaca/wool lined. "Converto-collar" opens into a hood, detachable neck-tab. Gray tweed. $19.67. [$25-30] **Suburban Jacket in 100% Pure Wool Tweed.** Donegal-type tweed "looks smart over everything. Well-tailored jacket borrows its handsome styling from the boys..." Simulated side vents and change pocket, two real pockets. Rayon twill quilted lining. Brown tweed. $19.67. [$25-30] Fall/Winter 1957.

Assorted casual spring jackets. The striped jacket features a popular feature, the convertible collar. $3.77-$11.74. [$20-25] Spring/Summer 1958.

Spring short jackets in a variety of fabrics and styles. **Washable Jackets.** Nylon fleece, cotton poplin and cotton sateen. $4.44-$5.87. [$15-25] Spring/Summer 1959.

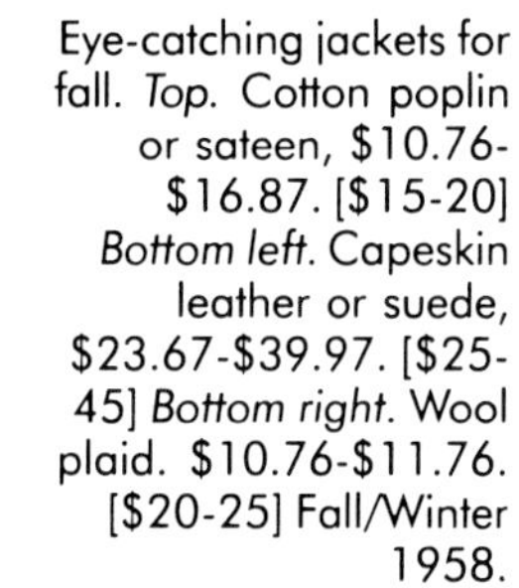

Eye-catching jackets for fall. *Top.* Cotton poplin or sateen, $10.76-$16.87. [$15-20] *Bottom left.* Capeskin leather or suede, $23.67-$39.97. [$25-45] *Bottom right.* Wool plaid. $10.76-$11.76. [$20-25] Fall/Winter 1958.

Mitin® mothproofed. Unlined. Plaid or solid. Red/black/gray plaid, $11.74. [$20-25] Solid in bright red, $10.74. [$25-30] Fall/Winter 1957.

Dry Clean Jackets. Corduroy, water repellent cotton poplin, and cotton knit. $6.77-$8.71. [$15-20] Spring/Summer 1959.

His reversible jacket is water repellent cotton poplin on one side, wide-wale corduroy on the reverse. Tan poplin, brown corduroy. $15.70. [$40-45] Her casual jacket is of water-repellent Parka® cotton poplin with corduroy trim and lining. Medium blue. $12.84. [$15-20] Fall/Winter 1959.

Orlon pile-lined jackets in cotton poplin, cotton sateen, or corduroy. $12.84-$16.97. [$20-25] Fall/Winter 1959.

Leather jackets in topper length, classic belted, and classic shawl collar button front styles. In suede or capeskin leather. $23.70-$39.97. [$30-45] Fall/Winter 1959.

All-weather coats in beautiful paisley, tiny checks, and floral tapestry designs. *From left.* Water-repellent rayon twill, acetate taffeta, and cotton hopsacking. The aqua blue and white check coat includes a matching umbrella and hat. From $12.70-$19.70. [$25-35] Spring/Summer 1959.

Sears own "Continental" Rainwear, styled by European designers Serge Kogan of Paris and Princess Giovanelli of Italy but tailored in America. Kogan's acetate twill lined in acetate taffeta, black with red lining. Giovanelli's iridescent Celaperm (acetate) and cotton, lined in acetate taffeta, helmet-type self lined kerchief, mocha. $19.67 each. [$95-115] Fall/Winter 1958.

Luxurious Washable Orlon Fleece Topper. Featherlike, wrinkle-resistant, mothproof, mildewproof, nylon lining. $17.40. [$65-85] Fall/Winter 1957.

The "Shorter-Than-Long" Coat. Three far-spaced buttons. Coat drapes to an oval in back, adjustable cuffs, acetate satin lined. Ruby Red. $24.50. [$55-65] **Oval Silhouette Wool Coat.** Draped back, low button front, big collar. Sapphire Blue. $49.50. [$85-110] **Black Looped Tweed Coat.** Rayon and acetate crepe-back satin lining. $39.50. [$65-75] Fall/Winter 1958.

Double Breasted Plaid Wool Coat. Lined in acetate and rayon crepe-back satin. Blue, green and black. $44.50. [$45-50] Fall/Winter 1958.

Rippled Textured 100% Wool Chinchilla Cloth. Box pleat and paneled back, acetate satin lined. Bright red. $36.50. [$45-50] **Lustrous Velour Coat.** Wool, in popular oval silhouette shape, acetate satin lined. Rose taupe. $32.50. [$75-95] **Low Belted Wool Donegal Type Tweed.** Acetate satin lined. Dark brown and white. $32.50. [$45-55] Fall/Winter 1958.

Oval Silhouette Velvety Wool Plush. Relaxed cowl collar, acetate and rayon satin lined. Sapphire blue. $44.50. [$65-75] Fall/Winter 1958.

Top left and clockwise. Beautiful coats in wool fleck tweed, wool/camel's hair blend, leather, polished wool zibeline, cotton corduroy, and all wool nub tweed. All lined with fur-like rayon or alpaca pile. $24.67-$64.50. [$45-65] Fall/Winter 1958.

Leather Chemise Suede Jacket. Pearl buttons, rayon twill lined. Rust. $24.97. [$45-50] **Wool Tweed Plaid Skirt.** Rust and gold. $7.97. [$15-20] **Bold Plaid Coat.** Wool/nylon/ cashmere blend. Rust, tan and black, black lining. $36.50. [$45-50] Fall/Winter 1958.

His and Hers Car Coats. Combed cotton with raccoon collar, fully lined in Orlon pile. Beige. Woman's, $19.97. [$20-25] Man's, $22.50. [$40-45] *Right.* **Corduroy Coat with Raccoon Collar.** Lined in Orlon pile. Water repellent. Deep beige, beige pile. $29.90. [$45-55] Fall/Winter 1959.

The Royal Dutch brand, named after the Dutch airline KLM, is the finest Mongolian cashmeıe tailored by Sears into luxurious coats. **Classic Clutch Coat** with stitched collar and front in light fawn brown, $79.00. [$95-125] **Fur Trim – Two Skin Mink Collar Coat.** Beige, honey blonde mink. $129.00. [$95-125] Fall/Winter 1959.

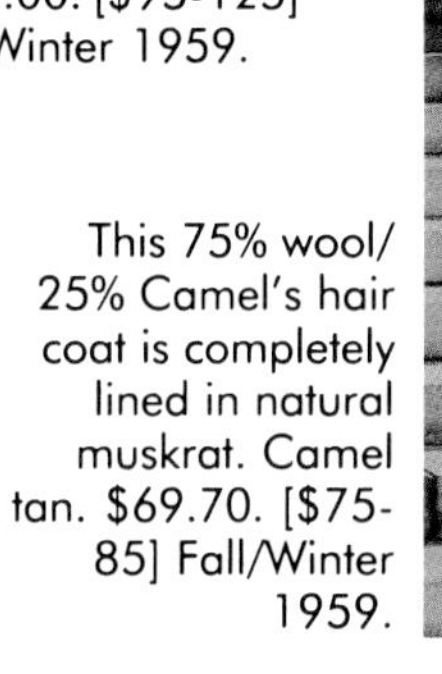

This 75% wool/25% Camel's hair coat is completely lined in natural muskrat. Camel tan. $69.70. [$75-85] Fall/Winter 1959.

The Boy-Coat, an exclusive, is styled from 100% imported menswear camel's hair. Lined in Milium insulated acetate satin. In camel tan. $49.00. [$95-135] A less expensive version in wool/camel's hair blend, lined in acetate satin. $29.00. [$45-50] Fall/Winter 1959.

The "Newest Looking Topper" is wool/camel's hair blend with a raccoon collar, lined in Orlon pile. Camel tan, beige pile. $19.90. [$45-50] **Raccoon Collar Poplin Long Coat.** Water repellent cotton, lined with Orlon pile. Sleeves lined with wool quilted to acetate taffeta. Loden green, black pile. $24.90. [$45-50] Fall/Winter 1959.

Beautiful and luxurious coats trimmed in real fur. *From left.* **Three-skin Collar of Brown-dyed Fitch (fur origin: Russia)** trims this 75% wool/25% mohair plush coat. Lined with acetate and rayon crepe-back satin. Light beige. $69.70. **Natural Ranch Mink Trim Collar.** 80% wool/20% fur fibers coat, lined with acetate satin. Black. $44.70. [$65-75] Fall/Winter 1959.

Sears finest Donegal-type wool tweed topper and coat. Convertible collar, detachable half belt. In brown, black, and white. Topper, $26.70. [$35-40] Coat, $32.70. [$45-50] Fall/Winter 1959.

Smart looking tweeds in looped tweed, check, and plaid styles. The left coat sports a rayon velvet trim. All completely lined. $29.90-$39.90. [$45-50] Fall/Winter 1959.

Economical but luxurious looking topper and clutch coat made of leather-like viny plastic. Both coats are lined in rayon pile. Black, white, or medium blue. Topper, $17.70. [$35-40] Coat, $19.70. [$40-45] Fall/Winter 1959.

The glamorous wide collar on this coat converts to a practical hood. In polished all wool Zibeline, available in long (shown) and topper length, available in black or medium blue, and in medium beige or moss green corduroy. All coats lined in rayon or Orlon pile. $19.70-$39.70. [$65-70] Fall/Winter 1959.

Odyna Solid Color Topper. Insulated acetate satin lining. In Pearly white, Platinum gray, Mist gray or Blond beige. $49.50. [$45-50] **Striped Short Coat.** Tones of Silver gray, Ranch brown, or Mist gray. $69.50. [$55-60] Fall/Winter 1957.

Sears Own Odyna Fur-Like Striped Coat. "*Odyna* is mothproof, mildewproof, odorless, exquisitely warm, light as a feather and it's treated for water repellency." Insulated acetate satin lining. In tones of silver gray, ranch brown, or mist gray. $79.50. [$45-50] Fall/Winter 1957.

Women's Outerwear. "Honestly, it Isn't MINK, It's Candalon® Wink, a man-made fabric, incredibly creating the look and feel of real mink...Blow upon its surface and see, as in true mink, the radiant ripple of color underneath." Made of acrylic and polyanide fibers on cotton, the Greatcoat and Stole must be furrier cleaned. Silver haze stole, $43.00. [$15-20] Amber glow greatcoat. $127.00. [$75-85] Fall/Winter 1958.

Notice that the same coat and stole offered above has come down in price from the year before. *From left.* **The Great Coat.** Amber glow. $79.90. [$75-85] **The Stole.** Caped back, cowl collar. Silver haze. $34.70. [$15-20] *Right.* **Orlon and Dynel Pile Coat.** The hood and convertible collar and cuffs are trimmed with mink-like pile. White with dark brown trim. $59.70. [$85-110] Fall/Winter 1959.

Teen Girls' Fashions

Dresses

By the end of the 1950s, separate sections in the Sears catalog were specifically designated for teen girls' fashions. This was the beginning of a marketing strategy aimed at the ever-growing post-war baby boom generation.

This was an era when girls wore dresses to school. The variety of fashions available gave a young teen many choices of dresses, not only for school wear, but for dating and parties. There are slim sheaths and full swing skirts, all with conservative necklines and the standard just below-the-knee length. For going out, a hat, bowclip or headband was added, along with a pair of short gloves and a pair of heels.

Teen 2-Piece Dress. Scalloped neckline cotton dress, print repeated in trim on cotton knit sweater. Yellow sweater with print on black. Set $6.44. [$45-50] **Sissy Shirtwaist.** Dan River woven plaid gingham, tie at neck, embroidery edged nylon ruffles, full skirt. Gray and red. $7.44. [$30-35] **Provincial Print.** Rick-rack trim on stand-up collar, elbow length sleeves and dirndl skirt. "Fashion's newest midriff-sleek line above a tiny-belted waist." Back zip. Amber gold print. $4.77. [$35-40] **Jacket and Dress Mates.** Full skirted sleeveless print dress, white pilgrim-style collar with bow. Black corduroy jacket buttons in back. Black and red print. $8.70. [$45-50] Fall/Winter 1957.

Swing 'n swirl full circle cotton dress in a geometric stripe print has 6 yards of swirl. Bright red and black. $5.90. [$45-55] Fall/Winter 1958.

Top to bottom, From left. **Leno Weave Checked Dress.** Square neckline, long line bodice and full skirt, fabric belt with flower. Combed cotton. Sea aqua blue and white checks. $9.54. [$35-40] **Lattice Yoke Cotton Dress.** Bead trimmed front yoke, deep scooped back. Sea blue print. $9.74. [$40-45] **Tucked Bodice Shirtwaist.** Cummerbund has crocheted looping. Dacron batiste. Sun yellow. $10.74. [$30-35] **Ruffled Coat Dress with Contrasting Long Sash.** Dark print cotton. Spice brown with sun gold flower print on black. $7.74. [$35-40] **Floral Cotton Dress and Sweater Set.** Grosgrain belt, Orlon knit sweater with applique roses. Set, $9.54. [$55-60] **Slim Dress and Button-Back Jacket.** Combed pettipoint cotton pique dress. Jacket with striped grosgrain ribbon. Copen blue and white. Set, $8.74. [$55-60] Spring/Summer 1958.

Assorted cotton dresses with generous use of sashes and full skirts. $4.76-$6.74. [$35-45] Spring/Summer 1958.

The "young version" of the classic shirtwaist dress has cluster pleats in the skirt, and a convertible collar. In acetate crepe paisley or Arnel and rayon broadcloth solid. Moss green, blue and pink paisley. Rose pink solid. $12.84 each. [$25-30] Fall/Winter 1959.

Two-way Chemise Dresses. To be worn belted or unbelted. Button down version in cotton/rayon with cotton velveteen bow. Rose or toast. Pleated back dress has detachable silk scarf. Navy or bright red, $6.94 each. [$35-40] *Right.* **Sheath and Bouffant Styles.** Blouson Jacketed Slim Sheath, knit cotton. $8.94. [$35-40] Ombre Box Plaid with black velveteen trim. $6.77. [$40-45] Fall/Winter 1958.

Buttons and Bows. White-edged bands trim. Detachable collar. Cotton. Black with white. $8.94. [$40-45] **Bold Woven Plaid Shirtwaist.** Criss-cross ribbon tabs, glitter buttons. Cotton. Red and mist gray. $5.84. [$30-35] **Set-in Cummerbund Check.** Dacron/cotton blend. Linen-look rayon lace-edged collar. Black and white. $10.84. [$45-50] **Brushed Flannel Casual.** Rayon and acetate with two-tone herringbone weave. Bowed belt. Light teal blue. $7.94. [$30-35] Fall/Winter 1959.

Chinchilla Cloth Coat. Ripple textured 100% sheep's wool. Lined in insulated acetate satin. Bright red. $32.70. [$75-80] **Two-piece Peplum Dress.** Wool flannel with matching rayon satin piping, fabric buttons down back. Light spice gold. $10.84. [$20-25] **The Junior Basic Dress.** Fine wool jersey, lined from waist to below hips. Moss green. $9.84. [$20-25] Fall/Winter 1959.

Designer Mary Lewis teamed up with Sears to "bring outstanding dress values at the lowest prices." **Slim Casual Sheath.** Rayon and acetate gabardine tailored dress with scarf collar. Peacock blue with navy. $9.54. [$65-75] **Striped Full Skirted Dress.** Acetate woven with white overstripe, white faille collar. Patent-like belt. Dusty blue, gray, black and white. $10.84. [$45-55] Fall/Winter 1959.

Pure wool dresses for the fall wardrobe. **Scarfed Jersey Sheath.** Fringed scarf drapes through shoulder loops. Lined to below hips. Bright red. $9.00. [$20-25] **The Flannel Princess Dress.** Detachable collar and cuffs in white linen-look rayon. Turquoise blue. $13.50. [$35-40] Fall/Winter 1959.

Gay Nineties Vestee over a Belted Sheath. Dress is finely ribbed acetate/rayon with white collar and cuffs of linen-look rayon. Banded midriff in front. Vestee is woven check acetate rayon. Black and white. $6.84. [$45-50] *Right.* **High Line Sheath with Jacket.** Dress has detachable cotton pique lace edged collar and jabot. Rayon and Acrilan blend. Heather gray. $13.50. [$50-55] **Box Bag.** Leather look plastic with a genuine leather lining. Mirror set in lid. Red. $3.47. [$55-60] **Matching Kid Pump.** $8.70 pair. [$30-35] Fall/Winter 1959.

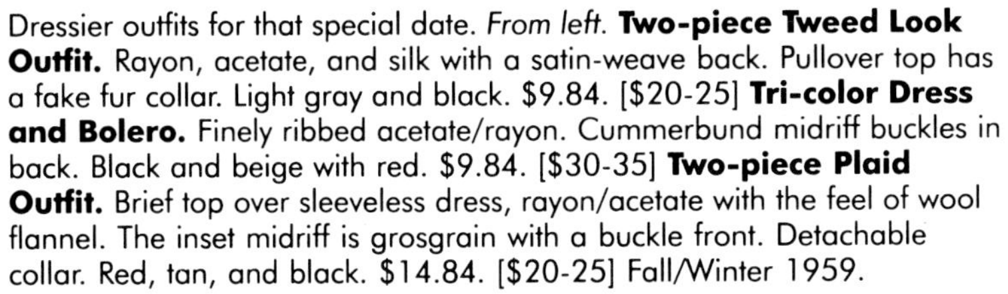

Dressier outfits for that special date. *From left.* **Two-piece Tweed Look Outfit.** Rayon, acetate, and silk with a satin-weave back. Pullover top has a fake fur collar. Light gray and black. $9.84. [$20-25] **Tri-color Dress and Bolero.** Finely ribbed acetate/rayon. Cummerbund midriff buckles in back. Black and beige with red. $9.84. [$30-35] **Two-piece Plaid Outfit.** Brief top over sleeveless dress, rayon/acetate with the feel of wool flannel. The inset midriff is grosgrain with a buckle front. Detachable collar. Red, tan, and black. $14.84. [$20-25] Fall/Winter 1959.

Dresses with the Kerry-Teen label featured classic sheath or bouffant styles. In rayon or cotton. $5.90-$8.97. [$25-30] Fall/Winter 1959.

Teen Girls' Fashions

Casual Separates

Young teen fashion separates include blouses, sweaters, skirts, jumpers, and slacks. Styles rarely differ from adult women's fashions, but are sized for the teen figure.

The following pages show a variety of teen casual fashions popular during the latter part of the 1950s.

From left to right, top to bottom. **Bulky Knit Orlon Turtleneck Pullover.** Rib knit on cuffs/bottom. Black. $6.77. [$15-20] **Virgin Orlon Cardigan.** Plastic shank buttons. Pink. $3.60. [$20-25] **Contrasting Collar Cardigan.** Angora mohair/wool collar, 3/4 raglan sleeves. Plastic shank buttons. Red with white collar. $3.94. [$35-40] **Shorty Cardigan.** 100% Orlon bulky knit, shank buttons with grosgrain ribbon backs. Red. $6.77. [$15-20] **Jacquard Ivy Leaguer** with all around knitted design. Bulky knit Orlon sweater. White with red and black design. $5.74. [$15-20] **Neckline News Pullover.** Virgin Orlon trimmed in Angora mohair and wool, loop-buttoned collar. Dutch blue. $2.94. [$35-40] Fall/Winter 1957.

Popcorn Stitch Bulky Cotton Knit Striped Cardigan. Pearly-buttoned panel. Slate gray, black and white. $3.94. [$35-40] **Stovepipe Hooded Pullover.** Combed cotton knit with heathertone stripes. Charoal gray and gold. $2.74. [$40-45] **Rib-Knit Bib-Front Turtleneck.** Corduroy-stitch striped cotton knit. Red and white. $2.83. [$25-30] Fall/Winter 1957.

Corduroy Princess Jumper. Square-neck fit-and-flare jumper. Button front panel. Bright peacock blue. $5.30. [$35-40] *All wool. $6.94. [$35-40]* **Corduroy Dutch Boy Skirt.** Detachable suspenders at hi-rise waist, unpressed pleats in front. Black. $4.77. [$25-30] Wool dacron blend.$5.40. [$35-40] Fall/Winter 1957.

Classic Orlon pullovers and cardigans. In white, pink, peacock blue, black, and amber tan. Pullover, $2.83. [$20-25] Cardigan, $3.77. [$25-30] **Orlon and Wool Pleated Skirts.** Orlon/wool blend, leather-like plastic belts. In three styles: **Sunburst Pleat,** fly-away flare plaid. In Blue or Red plaid. $5.94. [$15-20] **Double-Pleated,** slim lines. Solid Medium gray or Red and gray plaid. $4.77. [$15-20] **Ombre-Stripes** with swingy side pleats. Medium brown and gray stripe, or Blue and gray stripe. $4.94. [$20-25] Fall/Winter 1957.

Teen Personalized Dress. State 3 initials or *first* name on enclosed card and return with pocket tab. Bright red or Bright peacock blue. $5.90. [$20-25] **Bamboo Trimmed Bag.** Leather-like plastic, unlined. $2.46. **New Twosome Sissy Ruffled Princess Dress and Matching Stripe Sweater.** No-iron cotton dress, 100% wool bulky knit sweater. $11.60. [$45-50 set] Fall/Winter 1957.

Two-piece Knit Nubby-textured Suit. Angora rabbit hair trim on cardigan. Self belted skirt. Bright red. $9.60. [$65-70] **Wool Flannel Scoop-neck Jumper.** Buttons attach belt to front yoke, trim and skirt. Charcoal or Bright red. $6.94. [$20-25] **All-Wool Flannel Skirt.** Slim, high-rise waist, detachable suspenders have white saddle stitching. Black, gray and white checks. $5.67. [$30-35] **White Sissy Blouse.** Cotton broadcloth, black grosgrain ribbon tie. $2.27. [$30-35] **16-gore Felt Skirt.** Wool/rayon. "Really whirls!" Wide belt. Peacock blue or Red. $5.74. [$55-65] Fall/Winter 1957.

Coordinated outfits for mix and matching. Plaid is actually red and green cotton with solid colors in black, chino beige, or bright peacock blue. *From left.* **Chemise Style Vestee**, $2.97. [$20-25] **Pedal Pushers.** $2.57. [$35-40] **Sleeveless Blouse with Novelty Collar.** $2.34. **Capri Pants.** $2.97. [$45-50] **Jamaica Shorts.** $2.27. [$15-20] **Circular Skirt.** $4.77. **Short Sleeve Overblouse.** Side button closing. $2.97. **Slim Skirt.** $2.83. [$15-20] **Short Shorts.** $1.86. [$20-25] **Cotton Knit Sleeveless Pullover.** $1.57. [$15-20] **Shorty Pleated Skirt.** Attached bloomers. $2.97. **Coolie Hat.** Natural tan coconut straw braid with attached white scarf. $1.24. [$20-25] Spring/Summer 1959.

Sixteen-gore Felt Swing Skirt. Simulated leather belt. Black or peacock blue. $5.74. [$35-40] **Scoop-neck Pinwale Corduroy Jumper.** Pocket holds pure silk kerchief. Peacock blue, $5.94. [$20-25] Black and white check, $6.94. [$25-30] Fall/Winter 1958.

Kerry-Teen Orlon/wool skirts in assorted styles, and patterns. *Left to right.* **Sun-burst Pleated Plaid.** $5.94. **Reversible Pleated Skirt.** Light gray plaid on one side, red plaid on other, two-way zipper. $5.94. **Ombre-Stripe with Swingy Pleats.** Red stripe, blue stripe, or solid gray. $4.94. Fall/Winter 1958.

Teen girls' mix and match casual sportswear in brushed rayon plaid and solids. $2.97-$4.77. [$20-25] Full circle skirts in wool, wool/rayon blend, or corduroy. $5.94-$6.94. [$45-65] Fall/Winter 1959.

Ivy League Roll-Sleeve Shirt and Corded Cotton Slacks. Red plaid shirt, $1.97. [$15-20] Light blue slacks, $2.94. **Checkered Houndstooth Car Coat and Matching Slacks.** Beige poplin coat with checked corduroy trim, rayon lined. $9.90. [$45-65] Checked corduroy slacks, $4.60. **Sports Separates.** Assorted sportswear for the young miss. *Inset.* **Raincoats.** Red corduroy coat collar coverts to hood. $11.90. [$15-20] Flower print rubberized taffeta coat with coordinating cloche hat. Sky blue, black and white print. $6.60. [$20-25] Fall/Winter 1959.

Slim line teen fashions include the reversible wraparound skirt, jumpers, and separates. In various fabrics for fall. $4.54-$9.94. [$20-35] Fall/Winter 1959.

Classic sweaters for teens were similar to adult women's fashions. Priced from $2.77-$5.94. [$12-20] Fall/Winter 1959.

Teen Girls' Fashions

Lingerie

Since fashions in the late 1950s were usually form-fitting at the waist and hips, even teen girls needed "light control" in lingerie to smooth out the dress lines. Lingerie for teens included garter belt panties and briefs, and combination girdle/garter/brief because "busy teens love its freedom, its convenience." Bras featured the popular circular stitching favored by adult women. Slips and petticoats added flounce to the wide swing skirts. Panties and briefs were basically one style, high waisted and low at the legs. Most advertisements emphasized lingerie as a figure enhancing item of clothing to feel more "feminine,"rather than simply as undergarments.

The Bra Brigade. Made to accentuate the bust, these teen bras were fashioned to give unnatural "lifts." The rotation bra, fourth from left, promised its circular sun-burst will "give subtle accentuation to small busts." All bras in white. $.50-$1.86. [$15-20] Fall/Winter 1958.

The Girdle Chorus Line. Advertised as "perfect for full teen hips – gives comfortable, light control that teens enjoy," these undergarments ranged in price from $1.86-$3.77. Fall/Winter 1958.

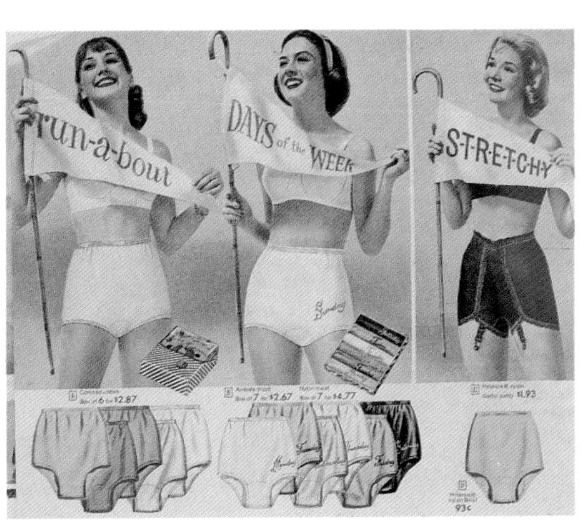

Underwear in cotton, nylon, or acetate tricot. In "soda pop" hues. Sold in box of 6 for regular briefs, and box of 7 for Days-of-the-week. $2.87-$4.77. [$15-25] Spring/Summer 1959.

Styles too pretty not to be seen included shaded floral flocked sheers, ribboned and lacy tiers, pastel multi-layer skirts. All petticoats in 100% nylon tricot. $2.94-$3.94. [$25-45] Fall/Winter 1958.

Bouffant petticoats for those whirly circular skirts. The pink petticoat offers a 50-yard sweep of frothy tiers of net. $2.97-$3.97. [$25-45] Fall/Winter 1959.

Teen Girls' Fashions

Fashion Accessories

The popular "poodle" theme appeared on everything from clothing to scarves to wallets. Knee socks and ankle socks were available to match sweater colors sold by Sears. Saddle shoes and different variations were popular for casual wear. They were so comfortable that high school cheerleaders wore them to lead cheers and do jumps. Athletic shoes for women and teens did not come into fashion until the 1960s.

Saddle shoes came in solid or two-tone styles. A popular choice for teen girls, these shoes were comfortable and stylish casual wear. Slight variations include back straps, and porthole trim. $4.77-$7.74. [$65-75] Spring/Summer 1958.

100% Virgin Orlon Blousette. In interlock knit with gray poodle applique, rhinestone eye. White, pink, or light blue. $1.97. [$45-65] **Jersey Poodle Stole.** 100% Virgin wool with fluffy yarn fringe, gray poodle applique with rhinestone eye, sparkling collar. White. $2.83. [$35-45] **Spun Rayon Challis Square.** For head or neckline, poodle print in one corner, ballet dancer in opposite corner. White or pink. $.93 or 2 for $1.74. [$35-45] **Clutch Bag and Wallet Set.** Leather-like plastic with painted poodle. Red, medium tan, light beige, or black. Set $1.50. [$35-45] Fall/Winter 1957.

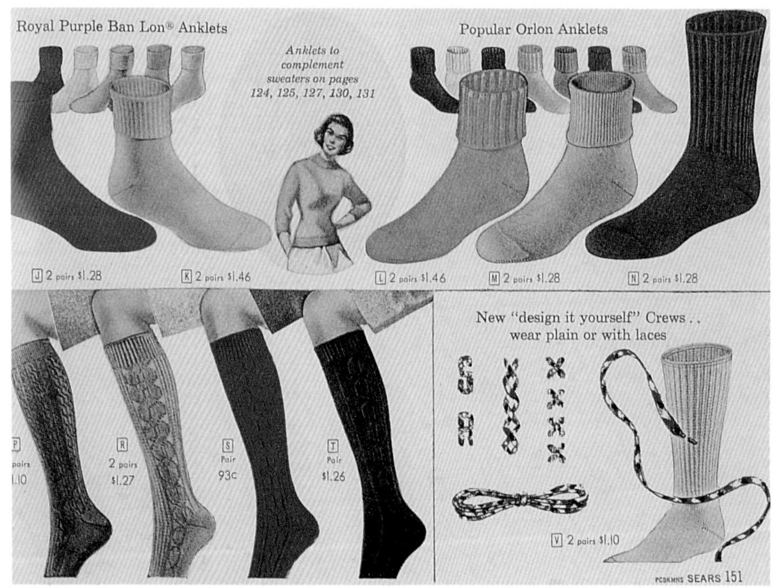

Assortment of socks including the new "design it yourself" crew socks with thread laces to personalize or create your own distinctive patterns. Two pairs for $1.10. [$12-15] Fall/Winter 1958.

Teen Girls' Fashions

Outerwear

In the late 1950s, fashionable outerwear for teen girls included jackets, car coats, and longer length "dress coats." Casual wear coats and jackets often featured the convertible collar that changed a collar into a convenient hood. Dress coats were long enough to cover skirts and dresses.

Double-Breasted Shawl Collar Jacket. Heavyweight cavalry twill cotton, 100% knit wool collar, rayon and acetate quilted lining. White, black trim. $13.70. [$25-30] **Wooden-Toggle Car Coat.** Collar converts to hood. Cotton sateen, striped rayon quilted lining. Red. $10.60. [$20-25] **Zipper Front Tyrolean.** Embroidery trim, draw-string hood. Cotton sateen, rayon and acetate quilted lining. Black. $11.74. [$55-65] **Washable Cotton Poplin Jacket.** Cotton flannel lining, elastic sides. Turquoise blue. $4.74. [$45-50] **Cotton Poplin Cap.** Wear ear flaps tied on top of head or down under chin. Self lining. One size fits all. $2.24. [$20-35] Fall/Winter 1957.

Teen Outerwear. *From left.* **Color-Flecked Splash Tweed.** Textured woven wool/nylon blend, rayon and acetate twill lining. Red Cotton velveteen trim. Belt tabs at side back. Dark blue and white. $19.64. **Striped Melton Cloth.** Wool/reprocessed wool blend, rayon acetate twill lined. Boxy style with tapered sleeves, flared back with inverted pleat and bow. Gray. $19.64. [$85-95] **Cardigan Coat.** Flaps conceal two real pockets, adjustable cuffs, belt tabs at side back. 100% wool fleece, rayon and acetate twill lining. Light beige. $23.64. [$65-75] Fall/Winter 1957.

Fall and winter jackets and car coats for teen girls. A popular feature was the convertible collar which turns into a hood. In cotton sateen, poplin, or nylon. $9.90-$12.60. [$25-35] Fall/Winter 1958.

A collection of jackets for teen girls in "the fashionable longer-length styles." The more expensive jackets are Orlon pile lined. $7.74-$14.90. [$25-45] Fall/Winter 1959.

Longer length coats for the "young miss." *Inset.* **Luxurious White Orlon Pile-lined Coat.** Completely lined wool/rayon tweed. Black and white. $23.90. [$35-45] *From lower left.* **Red Pebble-Weave.** Wool/nylon, lined in rayon and acetate. $17.60. [$35-45] **Rich Leather-Look Plastic.** 3/4 lined in black Orlon pile. Pearl white. $19.70. [$25-30] **Taupe Brown Orlon Pile.** Back contour belt, acetate lined. $29.90. [$25-30] **Corduroy Orlon Pile-lined.** Medium loden green thick 'n' thin wale, shawl collar converts to hood, fully lined with beige Orlon. $24.80. [$25-30] **Fur Collared Tweed.** Wool/nylon/cashmere tweed, dyed lamb collar (fur origin: Italy) has the rich appearance of blue fox. Rayon acetate twill lined. Royal blue and black. $24.80. [$30-35] **Two-Tone Braid Tweed.** Wool/rayon blend, rayon and acetate twill lined. Gray and red. $16.90. [$65-75] Fall/Winter 1959.

Smart Coats for

Men's Fashions

Suits and Dresswear

At first glance, men's suits, sports jackets, and slacks appear to remain unchanged through the decades. There are, however, many differences from year to year in lapel styles, buttons, tie widths, and trouser styles.

In the late 1950s, slacks were made roomier through the thigh and leg, and were cuffed at the hem. Slacks also were worn higher on the waistline. Suit jackets buttoned higher on the chest. Lapels, as well as neckties were beginning to be narrower - and would stay that way throughout most of the 1960s.

Of special interest are casual "campus-style" corduroy three-piece suits. It is hard to imagine today's college men wearing suits around campus, much less a "casual" suit.

Men's tailored dress trousers were double pleated, cut full and cuffed. *Left to right.* **Dacron/Rayon Decorated Sheen.** Charcoal gray. $8.40. [$65-85] **Orlon/Rayon Solid Tone Flannel.** "With the appearance of fine wool." Tan. $8.40. [$65-85] **Orlon/Rayon Fancy Fleck Pattern Flannel.** Medium blue. $8.40. [$65-85] Fall/Winter 1958.

Men's Suits. The premium quality striped dress suits were guaranteed for one full year against "rips, tears, or show undue signs of wear" or replaced for free. The year-round weight Dacron/rayon blend suits were considered wash and wear and could be machine dried. Popular 3-button style. In medium gray, medium blue, or medium brown. Notice the admiring secretary in the background. One trouser suit, $48.50. Two trouser suit, $63.00. Spring/Summer 1959.

Casual sport coats for summer leisure wear. *From left.* **Popular Miniature Pincord.** Washable cotton light weight. Blue. $9.60. [$55-65] **Galey & Lord "Zugara" Stripe.** Tropical weight Dacron/combed cotton blend. Wash and wear. Dark brown. $13.60. [$65-75] **Miniature Highland Plaid.** Tropical weight wash and wear, Dacron/combed cotton blend. Maroon plaid. $8.70. [$45-55] Spring/Summer 1959.

The "campus style corduroy suit" combines a unique coat, slack, and vest combination in a vertical wale corduroy. Reversible vest has a fancy rayon print that matches lining of jacket. Brass color buttons. Olive or tan. $24.50. [$65-75] Fall/Winter 1959.

Men's classic "Now..own your own..save on rental" formal wear. *From left.* **White Dinner Jacket.** Tropical weight rayon, rayon lined. $26.50. [$65-75] **Black Trousers.** Rayon satin braid trim. $11.50. [$35-40] **Sears Best Tuxedo Coat and Trousers.** Two-ply virgin wool tropical worsted, rayon satin shawl collar, rayon lining. $52.50. [$85-90] *Also in good quality rayon and Dacron. $39.50. [$65-70]* **Pleated Tuxedo Shirt.** White cotton, French cuffs. $4.94. [$20-25] **Rayon Cummerbund and Clip-on Bow Tie.** In black, midnight blue or maroon. $3.94. Fall/Winter 1959.

The man's check shirt also comes in a woman's version for that perfectly matched couple! The necktie has contrasting check trim to match shirt. Also pictured is an assortment of men's ready-tied Snapper® and Four-in-hand ties. Man's shirt, $3.86. [$12-15] Ties, $1.40 each [$5-15], any 2 for $2.76. [$10-30] Fall/Winter 1959.

Men's Fashions

Service Work Clothes

Men in service industry jobs were able to purchase work clothes from the Sears catalog. Not only were these clothes made out of materials tough enough to withstand heavy wear, but for a little more money, a woman could purchase for her husband or son the same items specially treated to resist dirt and grime. Naturally, the better quality clothing made wash day easier for her.

Men's Service Workclothes. Available in spruce green, army tan, airforce blue, gray-green, silver gray, charcoal gray, navy blue, or white. *Left.* **Chino Service Outfit.** Shirt, $3.24. [$25-30] Pants, $3.90 [$35-40], Special 3/shirts and 3/pants for $20.00. *Right.* **Army Twill Outfit.** Shirt, $3.24. [$25-30] Pants, $4.24. [$35-40] Outfit, $6.97. [$65-70] Fall/Winter 1957.

Army Chinos. Shirt, $3.12. [$25-30] Pants, $3.77. [$45-50]
Army Twills. Shirt, $2.90. [$25-30] Pants, $3.90. [$45-50]
Army Twill Outfit. Includes shirt and pants. $4.87. Cotton flannel-lined jacket, $5.70. [$55-75] *Unlined jacket, $3.77. [$45-65]* Fall/Winter 1957.

Men's Fashions

Casual Sportswear

Sportswear in the late 1950s was anything a man wore for leisure activity or casual occasions. Wearing a shirt and tie with a sweater or cardigan was considered casual wear.

A popular trend was the *Jac-Shirt*, which combined a shirt with a jacket into one piece. These shirts were convenient to wear and were worn outside the trousers. This combining of different elements was also evident in other items of clothing, including a combination shirt and vest, also in one piece.

Stripes and checks have always been fashion standards. During this period, plaids in different sizes and styles were especially popular.

Men's Sportswear. *Left to right.* **100% Wool Buffalo Overplaid Jac-Shirt.** $7.60. [$35-45] **Horizontal Corduroy Jac-Shirt.** $5.93. [$35-45] **100% Wool Overplaid Jac-Shirt.** $7.60. [$55-75] **Zipper Front Corduroy Jac-Shirt.** $4.93. [$55-75] Fall/Winter 1957.

Men's Sportswear. Jac-Shirts combine a shirt and a jacket that can be worn tucked inside trousers or outside. *From left.* **Blanket Stripe Cotton Flannel Pullover.** $3.76. [$35-45] **Kashara Cotton Flannel Jac-Shirt with Turtleneck Insert.** $3.76. [$35-45] **Heavy-weight Woven Flannel Ombre Plaid.** $3.93. [$45-55] **Popular Plaid Flannel Jac-Shirt.** $2.86. [$45-55] **"Newest Style Favorite" Blanket Stripe.** Sueded flannel. $2.86. [$55-65] Fall/Winter 1957.

Opposite page: **Nautical Knit Shirt.** Combed cotton with alternating jersey and mesh knit stripes on shirt. Men's, $3.80. [$35-40] Boys', $2.40. [$15-20] **Men's Beachcomber Pants.** White cotton gabardine with black and red stripes, tapered legs with side vents, zip front with extension-tab waistband. $4.80. [$45-55] **Boys' Surf Pants.** $2.80. [$20-25] Spring/Summer 1958.

Men's Swimwear. Cotton cabana sets in carribean print or "Martinique" stripe border. $6.86 set. [$55-65] Assorted swim trunks. Spring/Summer 1958.

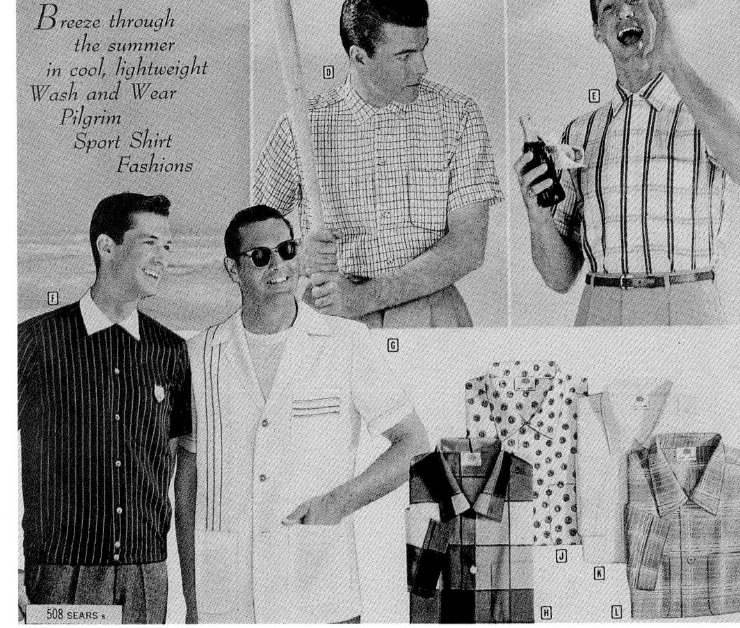

Assorted men's summer sportshirts. *Top.* **Nylon Plaid Shirts.** $2.90. [$20-25] *Bottom, From left.* **Navy Striped Jac-Shirt.** Dacron Cotton. $4.97. [$35-40] **"Gentleman Loafer" Shirt.** Dacron Cotton. $4.97. [$45-50] **Assorted Dacron Cotton Shirts.** In plaid, solid, circle paisley and woven plaid styles. $2.90-$3.86. [$20-25] Spring/Summer 1958.

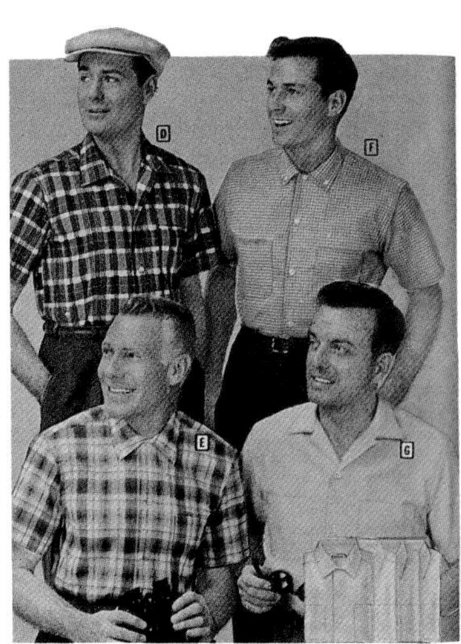

Assorted men's sportshirts in Dan River cotton. $1.90-$2.90. [$20-25] Spring/Summer 1958.

Dual-wear shirts button up for dress style and open-collar casual styles. Combed cotton. $2.90 each. [$20-25] Spring/Summer 1958.

Wearing a shirt and tie with a sweater or vest was considered casual wear. Men's sweaters in crew-neck, pullover, and button-down coat styles were available in hand washable 100% Hi-bulk Orlon and sold under the Pilgrim label. In camel tan, light blue, turquoise, cranberry, and charcoal gray. $4.70-$6.70. [$35-45] Fall/Winter 1958.

Fall and winter sweater styles featured cardigan vests and sweaters, pullovers, and Scandinavian-inspired jacquard patterns. In 100% wool or lamb's wool/orlon blend. $4.84-$9.97. [$45-50] Fall/Winter 1958.

Men's knit sportshirts in regular and continental flare collar style. In combed cotton, cotton knit, and pineapple mesh. $1.90-$4.90. [$35-55] Spring/Summer 1958.

Assorted men's sportshirt styles include shirts in textured acetate or wash and wear Dacron and cotton. $2.70-$3.86. [$35-65] Fall/Winter 1958.

Ivy-league style cotton shirts in check, stripes, foulard patterns, and plaids. $2.83-$3.86. [$15-20] Fall/Winter 1958.

Men's Jac-Shirts were designed for indoor and outdoor wear. In 100% nylon fleece or cotton flannel. $2.83-$7.84. [$45-65] *Second from right.* The one-piece shirt appears to be a 2-piece coordinated outfit. Fall/ Winter 1958.

Casual Jac-Shirts are 100% cotton with the light "outerwear look." Tailored to be worn outside the trousers. *From left.* **Nautical Motif.** Solid color collar and front placket. Two-button adjustable waist. $2.94. [$55-60] **Vest Style with Striped Collar Bib.** One pocket, adjustable button tabs on each side of waist. Tan hues. $2.94. [$55-60] **Combination Mesh Leisure Jac-Shirt.** Shirt front is cotton mesh and broadcloth, black and pockets are broadcloth. Black and white. $3.94. [$65-70] **Plaid Vestee Shirt.** Attached collar and front bib. Red plaid. $3.94. [$55-60] Spring/Summer 1959.

Convenient wash and wear cotton shirts for men in check, heraldic print, ombre panel, and madras plaid. $3.86. [$25-35] Fall/Winter 1959.

Men's fall sweaters. *From left.* **Jacquard Pattern.** Scandinavian influenced design, with shawl collar. Lamb's wool/ Orlon blend. Green. $8.97. [$35-40] **Boat-Neck Shaggy.** Brush "hairy" texture. Lamb's wool/Orlon blend. Gray heather. $6.90. [$25-30] **Continental Boat-Neck Big Stitch Knit.** 100% virgin wool worsted. Blue. $12.97. [$35-40] Fall/ Winter 1959.

Campus casual wear Jac-shirts combine the best style features of a lightweight jacket with a sport shirt. "Makes the gals look twice." *From left.* **Vest-Look Paisley.** Cotton broadcloth "shirt", acetate "vest." $3.86. [$55-60] **Sleek Vestee.** Rayon challis body and sleeves, acetate multicolor striped bib and collar. $2.86. [$55-60] **Concealed Button Front.** Bulk acetate in textured pattern, embroidered crest on pocket. $3.86. [$55-60] **Wash and Wear Cotton Vestee.** Woven plaid madras front, broadcloth bib, collar, sleeves and back. $3.86. [$55-60] Fall/Winter 1959.

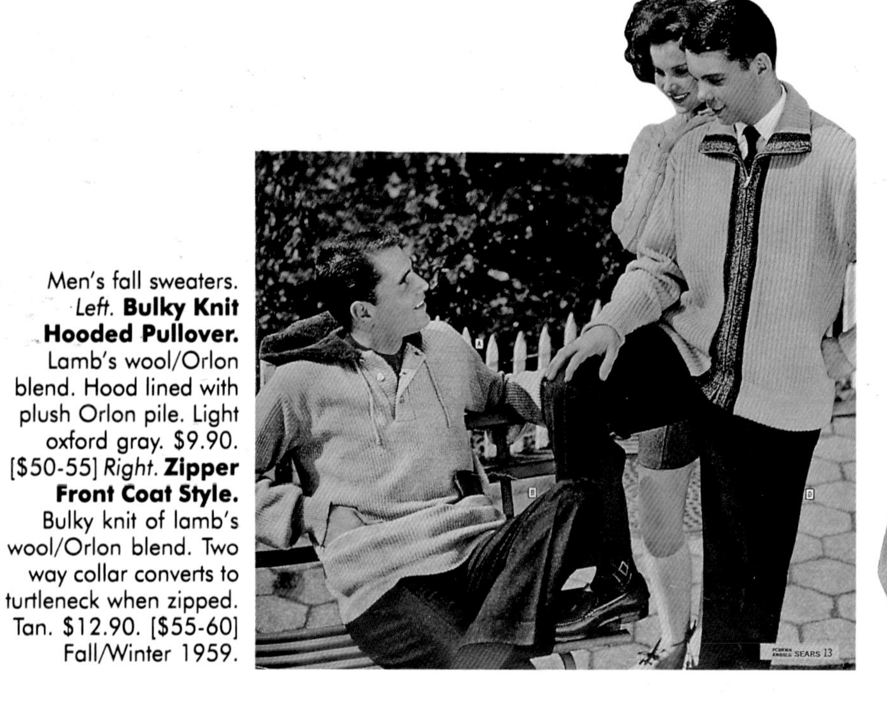

Men's fall sweaters. *Left.* **Bulky Knit Hooded Pullover.** Lamb's wool/Orlon blend. Hood lined with plush Orlon pile. Light oxford gray. $9.90. [$50-55] *Right.* **Zipper Front Coat Style.** Bulky knit of lamb's wool/Orlon blend. Two way collar converts to turtleneck when zipped. Tan. $12.90. [$55-60] Fall/Winter 1959.

The combination vest and shirt look in corduroy and Kashara cotton flannel. $3.86. [$55-60] Zip-front corduroy Jac-shirt in rust, $4.97. [$55-60] Fall/Winter 1959.

Men's Fashions

Sleepwear

Fashion styles for men's sleepwear were more varied than they are today. Ski-cuff style pajamas for winter were popular, especially for younger men. A wide variety of designs are represented here from the two-piece-look pajama top to a blazer style.

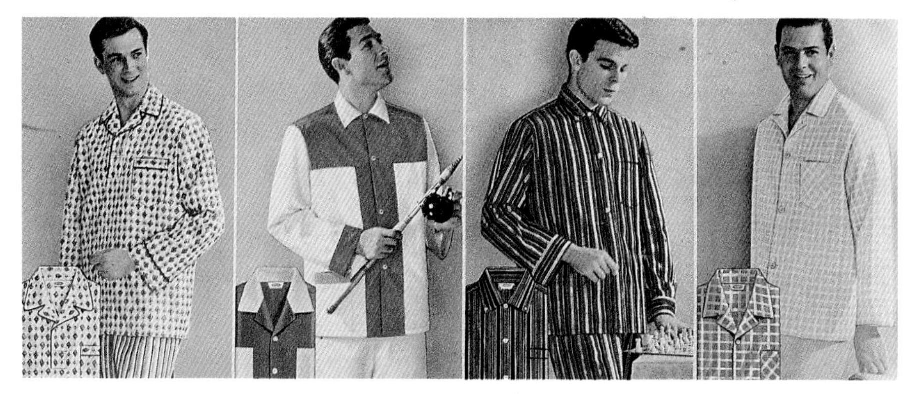

Men's Sleepwear and Loungewear. Men's sleepwear pants had elastic boxer style waistbands while the loungewear pants had belted fronts and elastic back-waists. Various styles and designs in cotton broadcloth or cotton rib knit were priced from $3.77-$5.77. [$25-45] Fall/Winter 1958.

Cotton broadcloth men's pajamas. Note the cardigan shirt look and the stripe blazer jacket style lounger. $3.86-$4.86. [$35-55] Fall/Winter 1959.

Men's Fashions

Accessories - Hats & Shoes

Prior to the 1950s, a well-dressed man rarely appeared in public without wearing a hat. By the end of the decade, men's hats had pretty much disappeared from the scene. This was evident by the few pages in the Sears catalog advertising hats for men. In the late 1950s, Sears sold dress hats under its *Pilgrim* label, some of which are shown here.

Athletic shoes, which were non-existent for women during this period, were mostly made of canvas duck with rubber soles. These were worn primarily to play basketball and were referred to as "basketball shoes." Sears sold these styles as *Arch Jeepers*. Compared to today's athletic shoes, men and boys had few choices in style or color.

It is not surprising, however, that dress shoe styles for men have remained relatively unchanged through the decades since the 1950s. Shown here are some examples featuring novelty trims and flaps that obviously have not survived.

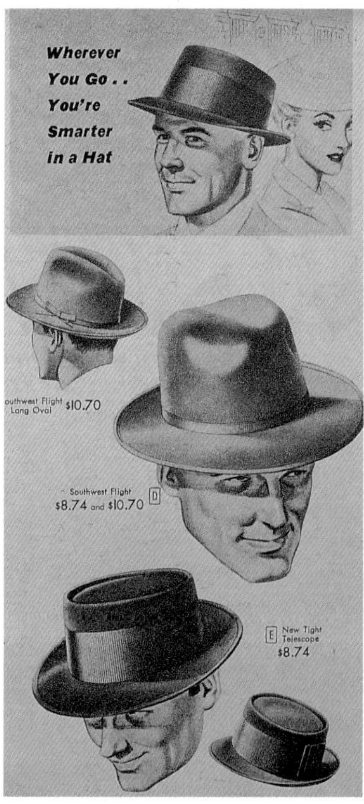

Assorted dress hats, priced at $8.74 [$45-50] for "better" quality fur felt and $10.70 [$45-50] for "best" quality fur felt. *Left to right.* **The Swagger, The Ambassador, Southwest Flight Long Oval, Southwest Flight,** and **New Tight Telescope.** Fall/Winter 1957.

Pilgrim brand dress hats for men in five popular styles. *Left.* **Jet Flight.** In two styles, regular and long oval. $3.90 [$45-55] for economy wool felt. $5.90 [$35-40] for water repellent fur felt. $9.90 [$45-55] for the best quality fur felt with "sable" hand rubbed finish. *Right.* **The Ambassador. The Tight Telescope. The Swagger. The Pork Pie.** $2.86-$9.90. [$45-55] Spring/Summer 1959.

Men's Athletic Shoes. Note the plain styles in white (or black) only. From $2.77-$6.77. [$25-45] Fall/Winter 1957.

Men and boys had very few choices in athletic footwear. These popular styles were called Jeepers (ankle top), or Jeeper Oxfords (low-top). Duck uppers with duck lining, with molded rubber soles. In white or black. $2.83-$4.77 pair. [$25-45] Spring/Summer 1959.

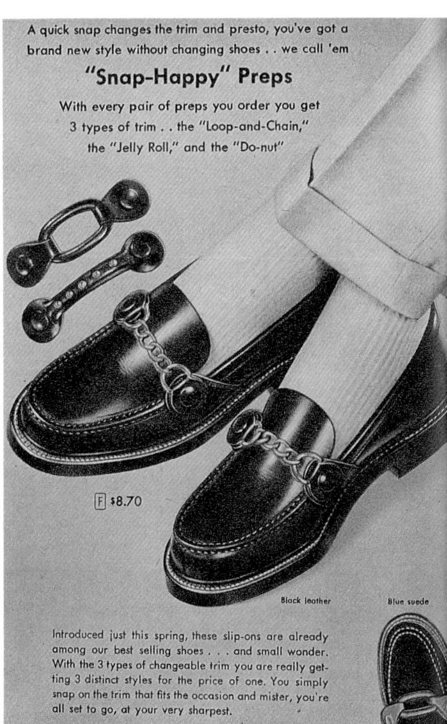

"Yes, sir, they're just about the smoothest thing in collegiate casuals." Leather or suede slip-ons with three types of trim. $8.70 pair. [$65-85] Fall/Winter 1957.

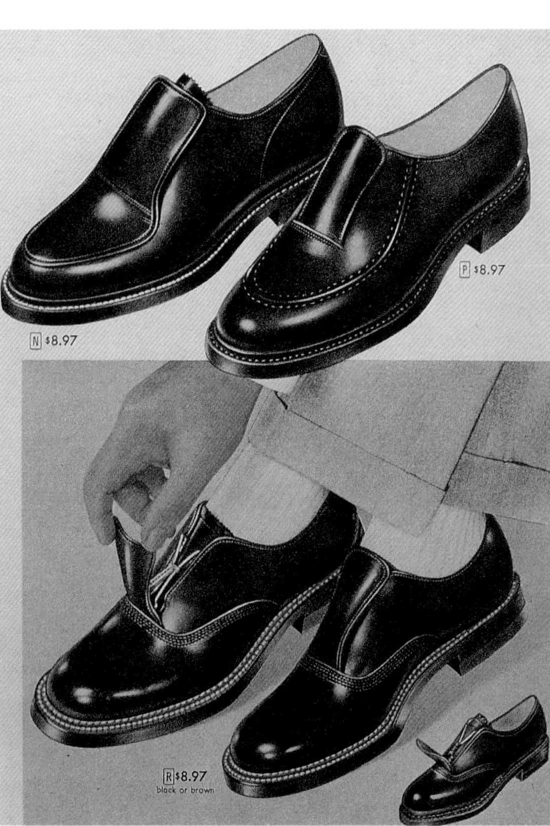

Dressy leather shoes featuring the Shu-Lok® fastener hidden under tongue. Three styles. $8.97 pair. [$75-95] Fall/Winter 1957.

Men's dress shoes have not changed much through the years. These shoes are made with rich calfskin leather with leather soles and rubber heels. From $13.97-$14.70 pair. [$65-85] Spring/Summer 1959.

Men's Fashions

Outerwear

Jacket and coat styles for men have not changed much since the late 1950s. Fabrics and fabric designs, however, have seen changes through the years. Notice the absence of jackets with professional sports team or college team logos that are so popular today. Shorter cropped almost to the waistline jackets have not survived the years.

Leather motorcycle jackets, sought after by vintage clothing collectors today, are pictured here virtually unchanged from 1957-1959. Even today's zippered leather motorcycle jackets have retained the timeless design popular in the 1950s.

The emergence of a new fastener, the *Velcro®*, was introduced in the late 1950s. This "radically new development in clothes fastening" revolutionalized the clothing industry.

Left and clockwise. **Sears Finest Leather Sport Coat.** Cleanable 3-button brushed suede, rayon satin lined. Sand tone tan. $38.50. [$85-95] **Suede Backstrap Cap**. $2.86. [$35-40] **Belted Zip-Front Leisure Coat.** Detachable two-piece belt. Sand tone tan. $29.50. [$65-75] **Three-button Suede Casual Coat.** Camel tan. $32.50. [$85-95] **Fieldmaster Western-style Suede Suburban Coat.** Made to wear over sport coat. Sand tan. $39.50. [$95-115] **Rancher Dress Western Fur Felt Hat.** Open crown, flat snap brim, grosgrain band, red acetate lined. $19.50. [$35-45] Fall/Winter 1957.

100% Wool Hooded Toggle Coat. Zip-on hood, quilted lining, concealed front zipper. Loden Green. $19.64. [$25-35] Fall/Winter 1957.

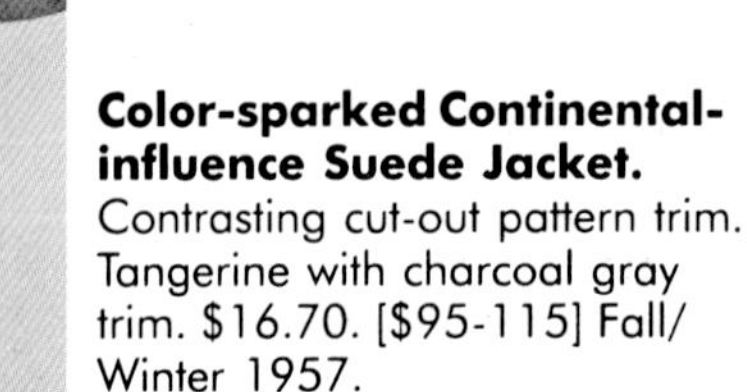

Color-sparked Continental-influence Suede Jacket. Contrasting cut-out pattern trim. Tangerine with charcoal gray trim. $16.70. [$95-115] Fall/Winter 1957.

Left. **Plaid Shirt Jacket.** Wool/nylon blend. $11.50. [$35-40] **Coordinating Corduroy Putter Pants.** $5.70. Set [$20-25], $15.50. [$55-65] *Top, left to right.* **Middleweight Fall Jackets: Blouse-type Wool/cashmere.** $10.90. [$85-115] **Reversible Woolen Plaid.** $9.70. [$65-75] **Rayon Sheen Gabardine Reversible.** $7.40. [$95-145] Fall/Winter 1957.

Assorted dressy quilt-lined tweed and fleece suburban coats. $19.50-$19.64. [$35-55] Fall/Winter 1957.

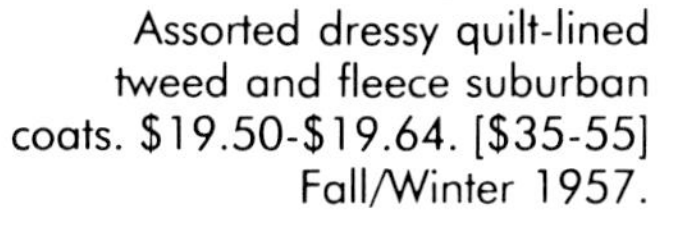

Top row. **Assorted Nylon Fleece-lined and Reversible Jackets.** $7.50-$13.50. *Bottom row.* **Washable Nylon Fleece Jackets.** Some quilt lined. 412.50-$19.50. [$65-75] Fall/Winter 1957.

A "radically new development in clothes fastening," VELCRO® front closures are used in this handsome cardigan and jacket. Rayon sheen cord cardigan is fully lined in rayon. Red. $9.70. [$35-40] Iridescent cotton and nylon jacket with contrasting cable knit trim, fully lined in rayon. Navy. $11.70. [$45-50] Spring/Summer 1959.

Assorted men's steerhide leather B-15 Bombers, motorcycle, and surcoats. $16.94-$39.50. [$195-300] Fall/Winter 1957.

Economy and quilted thermal-lined leather motorcycle jackets. Jet black. Economy, $20.87. [$195-300] Thermal-lined, $26.87. [$195-300] The prices for these same jackets in 1957, shown on far left, increased only a little more than one dollar in two years. Fall/Winter 1959..

From left.
The B-15 Bomber Style. Dyed lamb fur collar. $22.87. [$225-500] **Warm Sheep-Lined B-15.** Dyed lamb fur collar. $26.87. [$225-500] **Popular Waistline Style Thermal Lined.** $20.87. **Extra Long 31" Surcoat.** Wool quilted lined. Detachable belt. $27.87. [$300-550] **Warm Sheep Lined Surcoat.** Dyed lamb fur collar. Detachable belt. $39.50. [$300-550] All in cordovan brown.The same coats in 1957 are shown on left. Fall/Winter 1959.

Hercules Horsehide Cossack-style Jacket. Nylon lining guaranteed to last the life of jacket or relined for free. Cordovan brown. $26.97. [$200-250] **Hercules Motorcycle Jacket.** Quilted nylon lining guaranteed, removable lamb fur collar. Jet black. $31.50. [$300-550] Fall/Winter 1957.

Hercules Horsehide Surcoat. Quilted nylon lining, three-piece belt with detachable front sections. $34.50 Regular. [$300-550] This price remained unchanged since 1957. Fall/Winter 1959.

Sears best motorcycle jacket has slim tapered lines, and scuff resistant rich grained leather. Two collars, regular and snap-on of dyed mouton processed lamb fur. Lifetime lining guarantee. Jet black. $31.87. [$350-600] Motorcycle cap, $2.97. [$65-75] Fall/Winter 1959.

Children's Fashions

Young Children - Dress and Playwear

In the late 1950s, children's clothing were made for a specific occasion or purpose. Unlike current trends, during the 1950s there were definite differences between dress wear, school wear, and playwear. When children came home from church or school, they were expected to change into play clothes.

The catalog advertisements shown here include children's separates and dresswear, and clothes made for play. Many of the items here are brother-sister sets, similar styles for boys and girls. A large number of playwear made in a western style are shown in a separate section of Western Wear for the family.

For historical interest, samplings of halloween costumes are included.

Little Boys' Shirts and Slacks. *Left.* **Shirt, Slacks, and Belt Sets.** Cotton flannel shirts, corduroy slacks, self belt. Toast, charcoal gray, or navy. Set $3.97. [$55-60] Shirt only $1.67. Slacks only $2.67. [$15-20] *Right:* **Dress Shirts and Slacks.** Sold separately. Shirts from $1.67-$1.77 (includes tie). [$15-20] Slacks from $2.53 for cotton- $2.83 for corduroy. [$15-20] Fall/Winter 1957.

Children's Sweaters. *Left.* **Ban-Lon® Textured Nylon.** Pullover $2.83. [$15-20] Cardigan $3.77. [$20-25] *Right.* **Orlon Sweaters.** With imported embroidered and jewelled emblems, simulated rhinestones and pearls, and Norwegian jacquard patterns. $1.77-$3.77. [$25-30] Fall/Winter 1957.

Children's Playwear. Children's Corduroy Playwear. Drop-seat overalls, cotton flannel shirts, corduroy cuffed slacks. $1.15-$1.67 shirts. [$20-25] $1.87 slacks. [$15-20] $1.87 button overalls [$15-20], $2.97 for zip front overalls. [$20-25] Fall/Winter 1957.

Children's Playwear. Children's flannel lined corduroy coordinated playclothes. Cuffed corduroy boxer slacks, toggle-front and zip-front jackets, flannel shirts, and corduroy Caps. $1.47-$3.77. [$35-45] Fall/Winter 1957.

Children's Playwear. Assorted children's corduroy and cotton flannel everyday playwear. Overalls, playsuits, coveralls. $1.87-$5.74. [$20-35] Fall/Winter 1957.

Boys' Playwear. Three-piece nautical-style playsuit includes jacket, bib pants, and matching beret. Cotton gabardine. Navy blue with brass-colored buttons. $5.74. [$55-65] Spring/Summer 1958.

Children's Playwear. Brother and Sister playclothes in cotton flannel lined corduroy. In toast brown, red, navy, or copen blue. Jackets, $3.27. [$25-30] Boxer Longies with turn-down cuffs to prolong wear, $1.97. [$15-20] Shirts, $1.27. [$25-30] Boys' cap, $1.67. [$15-20] Girls' cap, $1.17. [$15-20] Fall/Winter 1959.

An assortment of Halloween costumes reflect popular children's themes of the period, including Lassie, Lil' Abner and Daisy Mae, Casper, and Popeye the Sailor Man. All costumes are rayon or rayon taffeta. $1.44-$3.77. [$45-65] Fall/Winter 1957.

Halloween costumes reflect children's interests of the time. Included are Zorro the man of mystery, Satellite Joe the man of tomorrow, Woody Woodpecker, Superman, and Bugs Bunny. Most clothing were made of rayon or rayon taffeta, with masks made of vinyl plastic. $1.89-$2.79. [$45-85] Fall/Winter 1958.

Children's Fashions

Boys' Sleepwear. Assorted styles define popular boys' pastimes and interests of the period. Jet plane, baseball uniform, rockets, western themes, and football. In combed cotton knit, cotton flannel, or Orlon/rayon blend. $1.97-$4.87 set with robes. [$20-35] Fall/Winter 1957.

Young Children - Sleepwear

Children's sleepwear, like Halloween costumes, reflect popular culture themes of the times. By the end of the 1950s, television and movies influenced sleepwear design. Children's pajamas followed sports themes, circus themes, western themes, and space exploration themes. There is even an official Sgt. Bilko pajamas for little boys.

Jet plane, baseball uniform, rockets, western themes, and football. In combed cotton knit, cotton flannel, or Orlon/rayon blend. $1.97-$4.87 set with robes. [$55-75] Fall/Winter 1957.

Children's Sleepwear. *Left to right.* Norwegian Border Print Ski-type pajamas, boys' football pajamas, Dungaree-style pajamas. Sanforized cotton flannelette. $1.87-$2.27. [$20-25] Fall/Winter 1957.

Snug 'n warm colorful pajamas in sports, western, and ski styles. All in combed cotton knit or flannelette. $1.97-$2.44. [$45-65] *Third from left.* **The Official Sgt. Bilko© 1958 CBS Inc. Pajamas.** "Perfect for viewing his favorite TV program", has woven sergeant chevrons and insignia. $2.57. [$75-85] Fall/Winter 1958.

Pajama for boys included "grown-up" styles, and "Junior Space Commander" and space ship print designs. In cotton knit or flannelette. $1.88-$2.17. [$45-65] Fall/Winter 1958.

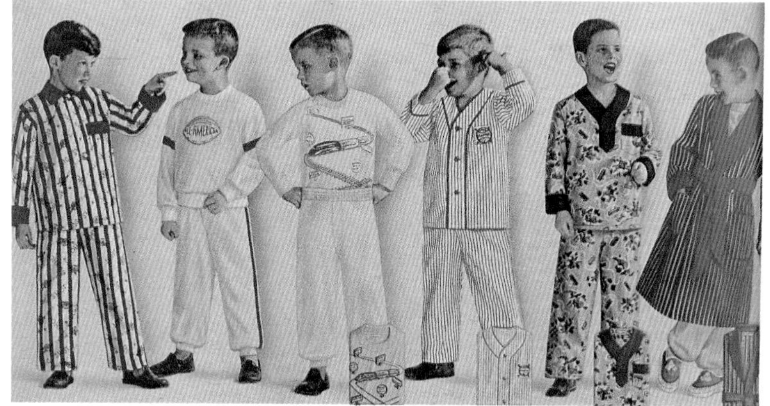

Designs feature "Big-Shot Cannon", All-American football, "Dreamland Express" chief engineer, Little Slugger baseball, the ever-present Western print, and the Ivy-stripe "Little Gentleman" robe. In cotton knit, flannelette, and suede cloth (robe). $1.87-$2.64. [$45-65] Fall/Winter 1958.

Three-Ring Circus pajamas in flannelette. **Clown.** Rick-rack and ruffle trim, pompon-concealed buttons. $1.97. [$25-30] **Kangajamas.** Toy kangaroo in pouch has its own kangaroo baby. $3.77. [$45-65] **Bedtime Bunny.** Puff cotton tail and toy carrot. $2.83. [$45-65] **Cat's Pajamas.** Including whiskers. For infants and children. $2.83. [$85-115] Fall/Winter 1959.

Western motif in flannelette. $1.87-$2.37. [$45-75] Fall/Winter 1959.

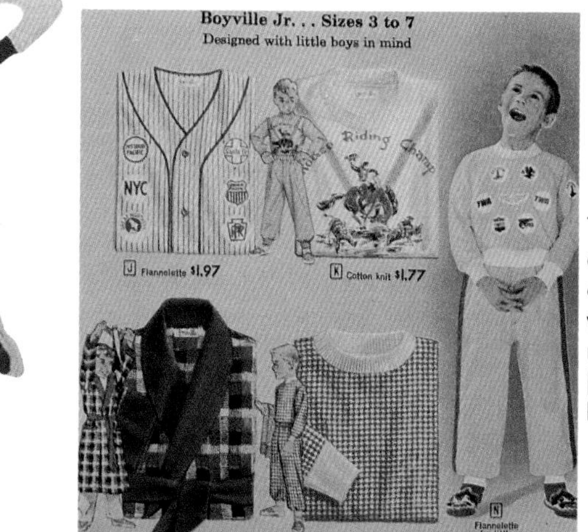

Themed boys' pajamas in cotton knit, cotton suede or flannelette. In baseball, western-style, or pilot motifs. $1.74-$2.64. [$45-65] Fall/Winter 1959.

Children's Fashions

Young Children - Outerwear

The collection of young children's jackets and coats are shown here in tweeds, nylon, fleece, and cotton. Many styles include matching hats or caps which usually tied under the chin. Some styles featured the "magic-grow" option that provided release hems in the sleeves and bottom to accommodate growth.

Top left and clockwise. **Three-piece Fleecy Coat Set.** Velveteen muff, poke bonnet. Turquoise blue. $10.76. [$65-70] **All Wool Fleece Coat.** Velveteen collar, rayon lined. Magic-Grow release hems. Matching bonnet. Pink. Set $12.72. [$20-25] **Washable Coat and Bonnet Set.** Nylon, trimmed in piping, quilted acetate lined. Light blue with navy. $12.72. [$20-25] **Brother-Sister Coat Set. Brother.** All wool check, adjustable hem. Ivy-league style hat. Medium blue with gray. Set $12.72. [$55-65] **Sister.** Velveteen-edged collar and sleeves trimmed with lamb. Velveteen covered buttons. Matching hat and tote bag. Set $12.72. [$55-60] **Splash Pattern Wool Melton Fitted Coat.** Cotton velveteen collar tie, matching piping set in waist, rayon lined. Matching bonnet has velveteen streamers. Charcoal with pink. Set $10.76. [$55-60] **Princess Style Coat.** Wool fleece, rayon pile collar and trim, cotton kasha lined. Matching bonnet trimmed with rayon pile. Red. $5.74. [$20-25] **Starlight Melton Coat and Hat Set.** 100% wool, cotton velveteen trim, rayon lined. Matching bonnet. Medium blue. Set $7.74. [$55-60] Fall/Winter 1957.

Lower left and clockwise. **All Wool Multicolor Tweed Coat and Hat Set.** Cotton velveteen collar, covered buttons and "pull-through" bow back trim. Tiny, $12.72. [$55-60] Small, $16.67. **All Wool Donegal-tyle Tweed Fitted Coat.** Back belt, lined with rayon. Tan. $12.72. [$45-55] **Clip Style Cotton Poke Bonnet.** Ponytail crown opening, rhinestone button trim. Beige. $2.14. [$20-25] **Textured All Wool Tweed Coat.** Black/white with red/white striped bulky-knit overcollar. Pull-through belt. $14,67. [$45-55] **Angora Rabbit Hair Cuddle Cap.** Rhinestone and simulated pearl trim, rayon lined. $1.87. [$15-20] **Splash-Pattern Magic-Grow Coat.** Release hems in sleeves and bottom. Knit trim on collar, pockets. 100% wool melton, rayon lining. $12.72. [$55-60] **Cotton Velveteen Clip Hat and Bag.** Set $3.41. [$30-35] Hat only, $1.87. **Fleece Play Coat.** Fleece collar and turn-back cuffs, quilted acetate lining. $11.74. [$55-60] **Basque Beret.** Imported from France, all wool felt with plaid knit band and earflaps. Red, navy, and black. $1.87. [$15-20] **Double Breasted Loose-style Coat.** Lined with bright corduroy, tassel trim on pockets and hood. Magic-Grow release hems. Royal blue with red. $10.76. [$55-60] **Budget-priced Loose-style Fleece Coat.** Wing collar trimmed with cotton plush. Rayon lined. Coral. $5.74. [$45-50] **All Wool Shaggy Felt Cloche.** Felt earflaps, trimmed with cotton velveteen studded with brass nailheads. White with navy. $2.34. [$15-20] Fall/Winter 1957.

Left panel, right to left, top to bottom. **Inverted Back Pleat Linen-Look Topper.** Rayon/nylon, rayon taffeta lining. Turquoise blue, sand, or pink. $3.77. [$30-45] **Pastel Nylon Fleece Topper.** Yoke-style back, half belts at sides, rayon taffeta lining. Pink, light blue or white. $6.77. [$30-45] **Washable Nylon Surcoat Jacket.** Tiny checked yoke front, Dynel collar, rib knit cuffs, lined in quilted acetate. Red or navy blue. $5.74. [$30-45] **100% Nylon Reversible Bomber-style Jacket.** Reverses from taffeta to napped fleece, bulky knit stand-up/turn-down collar. Red with reversible white. $6.77. [$30-45] **Cotton Sateen Tailored Jacket.** Fully lined with cotton sueded flannel. Red, turquoise blue, navy blue, or charcoal gray. $2.83. [$30-45] **Bomber-style Jacket.** Rayon/acetate sheen gabardine, Dacron quilted lining, Dynel collar looks like fur. Navy or gray. $5.74. [$45-60] *Right panel.* **Popular Car Coats.** New "Gro-Jac" features open seams in sleeves and bottom to increase jacket a full size. Water-repellent cotton sateen. Toggle closures. Girls and Boys sizes. $6.77. [$30-45] **100% Nylon Car Coat.** New "Gro-Jac" feature, nylon fleece lined. Red or turquoise blue. $8.71. [$30-45] Fall/Winter 1957.

Children's jackets and hooded parkas were limited to red, blue, and gray colors. The reversible styles proved a popular choice, two jackets in one. $5.74-$10.76. [$30-45] Fall/Winter 1958.

The classic tailored look in children's coats. In 100% wool, or wool/camel's hair blend. $13.67-$19.67. [$55-60] Fall/Winter 1959.

Children's Fashions

Girls - Dresses

Girls' dress styles were almost always fitted on top with full or flared skirts. Plaids and checks remained popular designs. Surprisingly, florals are almost absent. Many styles were available in different sizes for the big sister-little sister matched look. The "matador" look with little bolero jackets seemed to be a popular trend, especially in 1957. Toward the end of the 1950s, slim sheath styles started to appear.

Shirtwaister Dress. Gold cotton top with ¾ sleeves, braid trim. Multicolor woven plaid. Patent-like plastic belt. $3.94. [$15-20] **Poncho Outfit.** Woven plaid cotton with detachable adjustable side-lace poncho. Turquoise blue. $3.77. [$25-30] **Plaid Sweater Dress and Sweater Set.** Woven plaid gingham dress with patent-like plastic belt. 100% Hi-bulk Orlon cardigan sweater. Red with red, blue, and black plaid. $6.60. [$15-20] **"Sissy" Dress and Sweater Set.** Lace trimmed cotton dress with hi-rise midriff. 100% wool striped cardigan. Navy blue or light cherry red. $8.90. [$15-20] Fall/Winter 1957.

Woven Dan River Cotton Gingham Plaid. Black soutache braid and ball fringe, ¾ sleeves. Violet and rose plaid. $3.77. [$15-20] **Toreador Outfit.** Jacket with striped facing, dress with ruffled sissy bodice, detachable sequin tie, polished cotton skirt with pleated cummerbund effect. Black and white with red. $4.47. [$15-20] **2-Piece Plaid Woven Gingham Jacket Dress.** Dress with midriff bodice, cap sleeves and full skirt. Plaid trimmed jacket in cotton. Multicolor pastel plaid with light cocoa solid. $4.47. [$15-20] **Monogram Shirt Dress.** Cotton broadcloth with custom monogramming on detachable pocket tab. Bright red or peacock blue. $3.77. [$15-20] Fall/Winter 1957.

Tyrolean Lush Cotton Velveteen Jumper. Adjustable straps, braid rick-rack trim, acetate and rayon blouse. Red. Set $8.64. [$15-20] **2-Piece Set.** Cotton print dress with scallop neck. Pastel cotton knit cardigan with matching trim. Maize and black print with yellow sweater. $4.26. [$20-25] **Bolero-Jacket Set.** Polka-dot cotton dress with white collar and cuffs. Black corduroy bolero attaches to buttons that trim bodice. White dots on red. $4.47. [$20-25] **Acetate Taffeta Dress.** Full skirted with red nylon net "can-can" petticoat beneath, cotton velvet and lace trim. Navy and white check. $4.94. [$15-20] Fall/Winter 1957.

Girls' Big 'N' Little Sister Dresses. *Left.* **Dutch Boy Gingham Plaid Dress.** White waffle pique dutch boy collar, simulated pearl buttons. Small $4.57. [$20-25] Large $5.57. [$20-25] **Red Gingham Plaid Dress.** Red jersey top is completely lined, plaid trim on neck, bow, and cuffs. Black leather belt trimmed in gold-colored metal. Small $5.57. [$20-25] Large $6.67. [$20-25] Fall/Winter 1957.

Left: Girls' Big 'N' Little Sister Dresses. Slate gray cotton dress, white eyelet ruffle bib front, lace edged pique collar and cuffs, red bow at neck and sash at waist. Tiny $5.74. [$15-20] Small $6.57. [$15-20] Large $7.74. [$15-20] Fall/Winter 1957.

Right. Girls' Big 'N' Little Sister Coordinated Sets. Fitted style corduroy jumper with sweetheart neckline, pearl buttons. Cotton blouse with ruffles. Turquoise blue. Tiny set, $3.77. [$20-25] Small, $4.57. [$20-25] Large, $5.74. [$20-25] Fall/Winter 1957.

Little Girls' Dresses by Honeysuckle. Various styles featuring embroidered trim, Dutch-Boy collars, matador-style, pleated skirts. Fabrics including Acrilan jersey, combed cotton, cotton broadcloth and Orlon/rayon. From $3.87 [$15] for package of 3 - $5.57. Fall/Winter 1957.

Honeysuckle Fashions for Children. Big sister/little sister matching dresses, sister/brother sets, full circle skirts. Styles range from matador-style dresses to Tyrolean-style sets. $1.87-$4.77. [$15-25] Fall/Winter 1957.

Little Girls' Dresses by Honeysuckle. $2.57-$4.77. [$15-25] Fall/Winter 1957.

Navy and white dress clothes and matching accessories. *Left.* **100% Nylon Fleece Topper.** Nylon lined. $9.90. [$20-25] *Inset.* **Starched Cotton Lace Hat and Bag.** Set, $3.97. [$20-25] **Linen-Look Rayon Set.** $3.44. [$15-20] *Right.* **Blue Dress and Duster Ensemble.** Linen-Look Rayon. Set, $6.60. [$25-30] Spring/Summer 1958.

Pastel Spring Outfits. *Left.* **Acrilan Fleece Topper.** 6-gore back, rhinestones at petal collar, nylon lined. $10.90. [$45-55] *Inset.* **Three-Tier Daisy Toyo Band and Matching Bag.** Set, $3.97. [$20-25] *Right.* **Three-Piece Plaid Gingham Set.** Belted skirt, solid sissy lace blouse, and coordinated plaid blouse. Set, $4.97. [$20-25] Spring/Summer 1958.

Frosty Nylon Party Bouffant Dresses, all with separate rayon taffeta slip. Embroidered, or flock dot design. Pale aqua green, light blue, rose, lilac, or white. $5.70-$6.90. [$25-30] *Inset.* **Hat and Bag** set in embroidered organdy. White. $3.33. [$20-25] Spring/Summer 1958.

Left. **Pastel Checks in Various Styles**. $2.97-$3.97. [$20-25] *Top right.* **Two-piece Dress and Duster Sets.** $4.40-$6.60. [$25-30] *Lower right.* Little girl dresses that model adult styles. $3.40-$3.94. [$15-20] Spring/Summer 1959.

Plaid school dresses seemed to be a favorite for the younger girl. Coordinates to add versatility to a wardrobe included jumpers, ponchos, and removable and changeable dickeys. The full circle shirtdress has the look of a two-piece outfit. $3.77-$4.97. [$15-20] Fall/Winter 1959.

Girls' Honeylane cotton dresses for every occasion featured extras such as monogram embroidery, braid trim, schiffli embroidery, and matching bags. $3.77-$4.40. [$15-20] Fall/Winter 1959.

A selection of Honeylane full skirted school dresses and coordinated outfits. $3.77-$7.90. [$15-20] Fall/Winter 1959.

Children's Fashions

Girls - Casual Separates

School clothes in the late 1950s often consisted of blouses or sweaters, and skirts or jumpers. Slacks and shorts were only worn for recreational or leisure times. An assortment of casual separates are shown here by season and year.

Left panel. **Striped Blouse with Bow.** Multicolor cotton. $2.26. [$15-20] **Tabbed Corduroy Weskit Set.** Red. Set $4.77. [$20-25] **Tabbed Corduroy Jumper.** Reversible tabs on front yoke and pocket, reverses to match blouse. Peacock blue, or black. $4.26. [$20-25] *Top right.* **Mock Turtleneck Pullover.** $3.26. [$12-15] **Dutch Boy Pleats Corduroy Skirt.** High-rise skirt with detachable suspenders. Medium blue. $3.77. [$20-25] In wool and Dacron, $4.77. [$20-25] **Button-Up Corduroy Jumper.** Red. $3.77-$5.74. [$20-25] **Cotton Broadcloth Screen Print Blouse.** Matches skirt. White. $1.94. [$12-15] **Matching Felt Swing Skirt.** Wool/rayon blend. Screen print with glitter trim. Black. $3.47. [$25-30] *Lower right.* **Acrilan Jersey Pullover.** $2.94. [$15-20] **Sunburst Pleated Skirt.** Orlon/wool blend. $2.94. [$15-20] **Knife Pleat Slim Line Skirt.** Plastic belt. $3.77. [$10-12] **Reversible Pleated Orlon and Wool Skirt.** Light gray plaid reverses to navy plaid. 2-way zipper. $3.77. [$15-20] Fall/ Winter 1957.

Girls' Twin Sweater Sets**. Cable-Stitch Effect** 100% Orlon. Red. $4.47. [$15-20] **Mock-turtleneck.** 100% virgin Orlon pullover and cardigan. Peacock blue. Set $5.74. [$15-20] **Jacquard Design.** 100% virgin Orlon pullover and cardigan. Charcoal gray and white. $4.97. [$25-30] Fall/Winter 1957.

Left. **Bulky Hooded Blouson.** Zephyr wool. Red, black or white. $5.97. [$20-25 with cap] **Bulky Knit Turtleneck Pullover.** Textured weave in virgin Orlon. White or Navy or Red. $5.74. [$20-25 with cap] Matching Cuddle-Cap of Angora rabbit hair/nylon blend. Fall/Winter 1957.

Left. **Jacquard Design Bulky Pullover.** Virgin Orlon. White with red and black, or Red with white and black. $4.77. [$15-20] **Contrasting Stripe Collared Cardigan.** Virgin Orlon, glossy buttons. Bright copen blue. $3.77. [$15-20] Fall/ Winter 1957.

Two-piece Set. Floral print felt-textured black skirt, white cotton knit top trimmed to match. Set $4.44. [$25-30] **Jersey Blouse.** Contrasting stitch detail, wool string-bow at neck. White. $2.57. [$15-20] **Toggle-trimmed All Wool Suspender Skirt.** High-rise waist, soft front pleats, glittery "watch-fob" pin. Medium blue. $4.94. [$20-25] Fall/ Winter 1957.

Corduroy Car Coat and Pants. Collar on coat forms hood. Red or Black. $4.77. [$15-20] Pants has adjustable buckle tab at boxer back, slit-leg openings. Black or red. $2.26. [$15-20] **High Rise Corduroy Suspender Skirt.** Contrasting looping trim. Gold with black trim. $3.94. [$20-25] **Full Circle Corduroy Skirt.** Patent-like plastic belt. Medium blue. $3.30. [$20-25] Fall/Winter 1957.

High-Rise Quilted Cotton Suspender Skirt. Print backed with gold-color cotton. Detachable suspenders, plastic patent belt. Gold and gray. $3.94. [$20-25] **Blouse and Skirt Set.** Quilted print circular skirt with solid color backing. Matching blouse. Light peacock blue blouse with multicolor print black skirt. $5.20. [$20-25] Fall/Winter 1957.

Reversible Cotton Jumper. Navy blue on one side, polka dotted on reverse. Matching flock-dot blouse. Set, $6.94. [$20-25] Spring/Summer 1958.

Coordinated "Ivy-League Look" Clothing for big and little sister. Cotton. $1.47-$5.57. Spring/Summer 1958.

Girls Sun-time cotton coordinates in a bright yellow and green paisley print and glowing bright yellow shorts. Top, $1.97. [$12-15] Jamaica shorts, $2.27. [$12-15] Hat to match, $1.24. [$15-20] Spring/Summer 1959.

Completely reversible skirts and jumpers offer an economic way to have an expansive wardrobe. *From left.* **Pleated All Around Plaid.** Orlon/wool blend. Cranberry red plaid reverses to peacock blue and red plaid. $3.97. [$15-20] **Circular Skirt Side-Laced Jumper.** Solid red broadcloth reverses to red, white, and black plaid corduroy. $5.94. [$15-20] **Quilted Cotton Swing Skirt.** Solid blue reverses to blue print. $2.97. [$15-20] **Wrap-Around Plaid.** Braid trim, removable shank buttons. Wool/rayon royal blue and green plaid reverses to blue tweed. $3.97. [$15-20] Fall/Winter 1959.

Children's Fashions

Girls - Coats & Hats

Dress coats and casual jackets are shown here with a collection of spring hats and accessories. Once again, matching coats for big and little sister are offered. A popular design detail is the back pleats.

Girls' Big 'N' Little Sister Coats. **Washable sueded 100% Acrilan Topper.** Nylon taffeta lined, notched back collar and side belts. Light blue. Small $8.71. [$20-25] Large $10.76. [$20-25] **Black and White Loose Tweed Coats.** Wool/nylon blend, detachable bulky knit overcollar, inverted pleat and waist tab in back. Small $14.67. [$20-25] Large $17.67. [$20-25] Fall/Winter 1957.

Cardigan Dress Coat. 100% wool fleece, rayon/acetate taffeta lined. Turquoise blue/black trim. $17.60. [$45-50] **Spatter Tweed Dress Coat.** Wool/nylon blend, rayon and acetate lined, velveteen trim. Navy blue tweed, red trim. $16.64. [$50-55] Fall/Winter 1957.

100% Wool Plaid Dress Coat. Bow-centered yoke with rippling gored back. Rayon and acetate twill lined. Peacock blue and gray. $14.90. [$45-50] **Striped Melton Cloth Double-Breasted Princess Coat.** Wool/reprocessed wool, rayon and acetate twill lined. Genuine ermine tails and cotton velveteen accent. Charcool gray with medium gray. $19.64. [$35-40] Fall/Winter 1957.

Hats and bags for the well-dressed girl. Styles included bonnets, and mandarin hats, drum and parasol bags. $1.87- $3.40. [$15-25 hats, $65-85 bags] Spring/Summer 1958.

Girls' car coats in various styles. The newer, longer length jackets are featured in the inset. $6.97-$12.90. [$20-25] Fall/Winter 1959.

Children's Fashions

Boys - Dress and Sportswear

Clothing for teens and young boys included "prep" style sport coats and slacks, sweaters, and usually shirts with collars. Red, blue, and gray were popular colors during the late 1950s, and variations of these colors in different patterns and designs are shown here. Cuffed pants continued to be in style during this period, and were seen in both dress and casual style slacks. Sears marketed these items under the *Fraternity Prep* and *Boyville* labels.

Teen Boys' Fraternity Prep Sport Shirts. Assorted styles include plaid, stripes, Ivy-look checked trim, and embroidered westerns. Cotton broadcloth, chromespun acetate, combed woven cotton, and rayon "that feels like fine wool." Fall/Winter 1957.

Typical boys' shirt styles in combed cotton and cotton broadcloth. The embroidered western style shirts were available in "colors bright as a western sunset." An interesting variation is the one-piece vestee shirt that combined the look of a shirt and cardigan. $1.37-$2.87. [$15-20] Fall/Winter 1958.

All Wool Flannel Suit. $24.50. [$35-40] **All Wool Nubby Textured Sport Coat.** $16.50. [$20-25] *With flannel slacks, $22.95.* **Wool Flannel Fleece Sport Coat.** $14.95. [$35-40] *With gabardine slacks, $19.50. [$45-65]* **All Wool Sport Coat with Sparkled Nubs of Color.** $11.90. [$20-25] *With gabardine slacks, $16.50. [$45-65]* Fall/Winter 1957.

Sweater styles for boys in Orlon, Ban-Lon®, and Lamb's wool. Assorted boy's wash and wear knit shirts. $2.87-$3.84. [$15-25] Fall/Winter 1958.

Sweater styles fashioned after teen trends such as the award sweater. These items were labeled Fraternity Prep Knitwear, and were made of Orlon and wool. $2.83-$4.77. [$15-25] Fall/Winter 1958.

Boys' Ivy-League Outfits. Boyville outfits feature cuffed pants, and jackets. Available in cotton or denim, all flannel-lined. Shirt $1.67. [$15-20] Pants $2.97. [$20-25] Jackets $3.47. [$45-65] Fall/Winter 1957.

Matching outfits for little boys include flannel lined jackets, flannel shirt, and cotton gabardine or blue denim pants with matching cuffs. Shirt, $1.37-$1.47. [$15-20] Pants and suspender jeans, $2.27-$2.44. [$20-25] Jacket, $2.44-$3.27. [$45-65] Fall/Winter 1958.

Fire-engine red flannel lined denim and corduroy outfits. Shirt, $1.44. [$12-15]Jeans, $1.97-$2.47. [$45-65] Jacket, $2.47-$3.27. [$55-65] Fall/Winter 1959.

Children's Fashions

Sears Exclusive Woolen Surcoat. Fur-effect Dynel collar in random splash pattern. $8.65. [$55-75] **Sears Exclusive Duralon Surcoat.** Water repellent rayon/acetate/nylon blend in Weather-mist pattern. Leather reinforced cuffs, Dynel collar. $7.97. [$55-75] **Suburban Random Splash Pattern Coat.** Rayon quilted lining. $8.77. [$55-75] **Brushed Woolen Fleece Suburban Coat.** Rayon quilted lining, striped pattern. $10.95. [$45-50] Fall/Winter 1957.

Boys - Coats & Hats

Boys outerwear design are fashioned from men's styles. Motorcycle-style jackets, for practicality, came in plastic or leather. Reversible jackets were popular and practical, giving the advantage of two jackets in one. Included in this section is a selection of boys' rainwear, including an official *Dragnet* trench coat.

A variety of caps and hats are shown. Popular styles during the late 1950s included leather caps with turned-up ear flaps and a jet pilot style complete with aviation goggles.

Left. Practical wash and wear nylon parkas in nylon fleece or reversible taffeta and nylon fleece. $8.70-$10.97. [$35-40] *Right.* Woolen parkas. Orlon pile or quilt lined. $11.75-$13.65. [$35-40] Fall/Winter 1958.

Teen Boys Fraternity Prep Reversible Jackets. Cardigan style, checks, ivy-league style, and dress style. In cotton and nylon. $4.97-$9.67. [$35-45] Spring/Summer 1958.

Boyville Leather Jackets. B-15 Bomber-type, $12.75. [$150-225] Velvet-soft Suede Leather, $9.65. [$65-85] Scuff-Resistant Steerhide Leather Cycle Jacket. $12.75. [$185-250] **Sears Exclusive Bomber-style Jackets.** Weather Mist Duralon. Guaranteed one year against rips and tears. Leather reinforced cuffs. $6.79. [$65-85] Snug-fitting Bomber. $7.65.[$65-85] Fall/Winter 1957.

ubber Rain Set. Guaranteed not to crack, p, or scuff, or replaced free. Features "safety slogan" lining, water- repellent corduroy ollar. Matching hat. Yellow. $3.87. [$25-30] *Top row.* **Rubber Rain Set.** Black. $2.97. [$20-25] **Roy Rogers Western Rubber Rain Set.** Roy Rogers and Trigger on pockets. Black/yellow trim. $3.77. [$95-115] **Official Dragnet Trench Coat.** "Exact duplicate of coat worn by Jack Webb as TV's Sgt. Friday." Spun rayon with 100% water-proof plaid rubberized backing. Tan. $5.97. [$45-55] *Bottom row.* **4-way Plastic Hooded Poncho.** Converts to tent, sleeping bag, or ground sheet. Yellow. $3.97. [$35-50] **Roy Rogers Western-styled Set.** Includes plastic waterproof zippered courier case. Roy Rogers insignia and western fringe, Krene plastic. Sand tan/light tan. $2.88. [$95-115] **Slicker-styled Rain Set.** Krene plastic. Light tan or yellow. $2.88. [425-30] Fall/Winter 1957.

Boys Fraternity Prep Jackets in wash and wear tarpoon cloth paid or stripe, washable reversible cotton and poplin. $3.49-$5.97. [$25-35] The quilt-lined cycle jacket is waterproof scuff resistant plastic that looks like leather. $9.90. [$75-80] Fall/Winter 1959.

Lined Dress Jacket. Rayon-sheen gabardine accented with two-tone panel front. Dark blue. $5.70. [$95-150] **Sheepskin-lined Mackinaw.** Water repellent cotton moleskin cloth, lamb collar. Dark brown. $9.97. [$65-85] **Heavyweight Pea Coat.** Wool with scarlet rayon quilted lining. Navy inspired anchor design buttons. Navy blue. $8.98. [$65-75] Fall/Winter 1957.

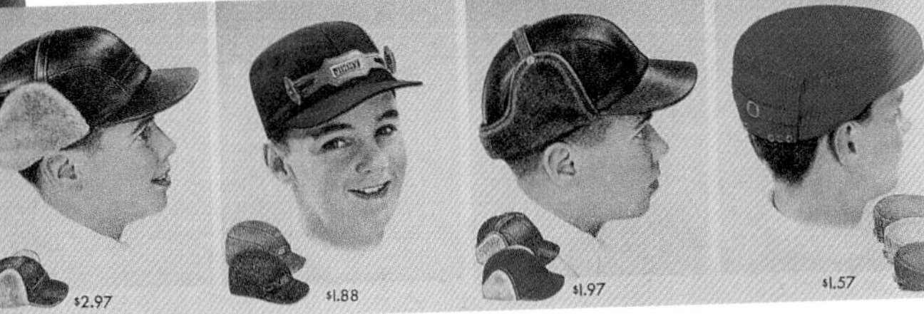

Jet pilot style with detachable goggles, knit skating style,and Roy Rogers Texas style. $.94-$2.39. [$15-65] Fall/Winter 1958.

Genuine Front-Quarter Steerhide Leather. Lamb fur turn down band. Brown. $2.97. [$30-35] **Personalized Cotton Sheen Gabardine Name Cap.** 100 boys' self-adhesive names included to attach. Navy. $1.88. [$25-30] **Capeskin Leather.** Alpaca wool ear flaps, cotton sueded lining, chin strap with snap fastener. Brown. $1.97. [$30-35] **Ivy League Corduroy.** Rayon satin lined. Beige-tan, charcoal gray, forest green, red, or rust. $1.57. [$25-30] Fall/Winter 1957.

Boys' Fraternity Prep leather and suede jackets including Sears finest black motorcycle jacket. $19.97. [$350-600] All other jackets are from $12.47-$19.97. [$75-95] Fall/Winter 1959.

Western Wear

In the late 1950s, western wear was so popular that fashionable clothing in that style for the entire family was available and sold through the Sears catalog. The clothing was geared predominantly to children, probably due to the popularity of television shows and movies in the western theme. Boys and girls could wear wonderful western wear from Sears sold in a generic brand, or jeans sold under the Sears *Roebucks* label, or entire outfits sold under the exclusive *Roy Rogers and Dale Evans* label.

Men's western wear consists mainly of shirts and hats. With the exception of western outfits for girls, women's western wear was non-existent except for a few examples of western-styled tooled leather handbags.

This trend continued into the mid-1960s when television westerns faded into the sunset.

Animated Buckin' Bronco Belt. Horse and rider in full color on buckle moves with every movement of body. Steerhide leather with tooled effect design. Black or tan. $.93. [$45-65] Fall/Winter 1957.

Authentic Western suits and accessories. **Roy Rogers 10-pc. Cowboy or Dale Evans Cowgirl Outfit.** Red and black corduroy and cotton suede-cloth pants, (Corduroy skirt with kick-pleat front), matching vest, cotton flannel shirt, leather belt, 2 leather holsters, 2 nickel-plated toy guns, scarf, and lariat. Set, $6.77. [$225-275] **Roy Rogers 9-pc. Cowboy Outfit or Dale Evans Cowgirl Outfit.** Cotton twill chap-front and maize pants, (cotton twill cowgirl skirt), cotton flannel shirt, leather belt, 2 leather holsters, 2 metal clicker pistols, scarf, lariat. Set, $4.77. [$225-275] **Roy Rogers 3-pc. Frontier or Dale Evans Cowgirl Set.** Embroidered polished cotton pants (skirt), jacket, belt. Red/black or maize/brown. Set, $5.75. **"Bonanza" Silvon Metalized Leather Holster Set.** Roy Rogers autograph, cap-firing 11" nickel-plated guns. Burgundy. $3.94. [$125-175] **Decorated Roy Rogers 1" Tanned Leather Belt,** with studs, simulated jewels, bullets, metal sheriff's badge, Roy Rogers autograph on patch. Black or tan. $.94. [$65-85] **Roy Rogers Rodeo Hat.** 100% wool felt, whip-laced brim, adjustable chin cord. Black, red or tan. $1.88. **Economy-priced Silvon Metalized Holster Set.** Cap-firing 8" guns, 4 silver color bullets, Roy Rogers autographed center plate. Silvon is imitation silver bonded to cowhide leather. Brown or black. $2.97. [$85-150] Fall/Winter 1957.

Children's Western Wear. Texas-made. Cotton twill slacks fully lined in cotton flannel. Matching flannel shirt with embroidered twill yoke front and back, western slash front pocket. **Western-styled pre-formed Hat.** Wool felt, rayon ribbon braid trim and gold-color stars around crown bottom. Adjustable cord drawstring. Buckskin Tan, Black, or Bright Red. $1.87. [$125-175] Fall/Winter 1957.

Roy Rogers and Dale Evans washable western wear were worn as everyday playclothes. Well made, the outfits were sturdy cotton twill and rayon gabardine, with fringe and embroidery trim. $4.77-$7.74. [$175-250] Fall/Winter 1958.

Generic label boys' western wear featured cotton flannel shirts, cotton twill or whipcord pants and flannel lined jackets. Hats sold separately. Shirt and pants set, $4.74. [$45-65] Jacket & pants set, $5.94. [$55-75] Fall/Winter 1958.

Coordinated western-style playwear in flannel-lined denim pants and coordinating shirts. $1.77-$2.83. [$45-85] Fall/Winter 1959.

Teens' Roy Rogers western wear jean/jacket outfits, "designed exclusively for the 'King of Cowboys' to be worn by all his young cowpokes." *Left.* **Jeans and Matching Rider Coat.** Vat-dyed blue denim. Roy's leather patch in back. $2.77 each. [$125-150] *Right.* **Cotton Flannel-lined Jeans and Matching Jacket.** $2.97 each, larger sizes $3.17. [$95-125] Fall/Winter 1957.

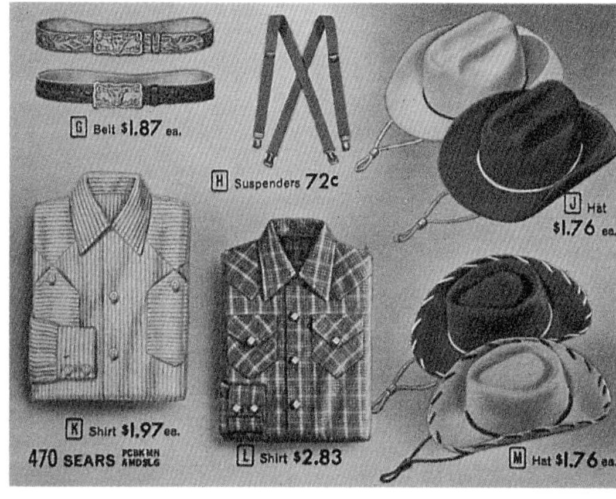

Western-style shirts and accessories for the junior ranger. $.72-$2.83. [$35-65] Fall/Winter 1959.

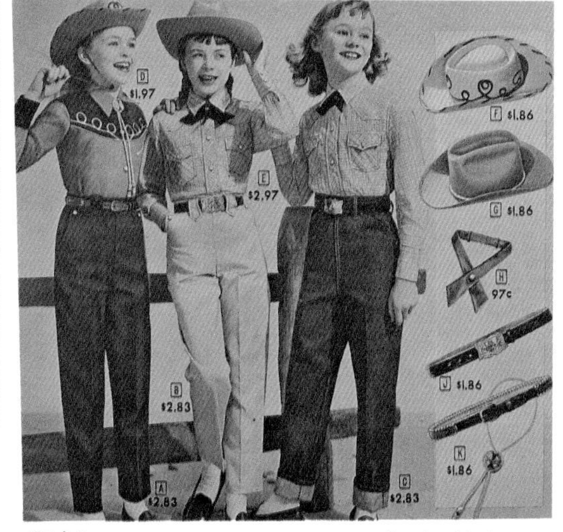

Girls' western sportswear accessorized with hats, crossover and cord string ties, tooled belts. In cotton or denim. Clothing from $1.97-$2.97. [$55-65] Accessories $.97 and $1.86. [$25-45] Fall/Winter 1959.

Roy Rogers 100% Orlon sweaters in jacquard and diamond knit, in western rodeo motif. $2.87-$3.84. [$95-115] Fall/Winter 1959.

Boys' zippered cotton stripe shirt, and the ever popular western style fringe frontier and double foldover styles. $1.37-$1.74. [$45-65] Fall/Winter 1959.

Foldover Front style shirt. $1.50. Roy Rogers Frontier Style with Fringe. Short or long sleeves. $1.37-$1.57. [$35-55] Fall/Winter 1957.

Men's Western Sport Shirts. **Embroidered Rayon Challis.** Piping trim, "arrow" pockets, "Caballero" cuffs. White, Turquoise, or black. $9.67. [$125-175] **"Gambler" Model Gabardine.** "Arizona" wing pockets. Light tan, wine, aqua, light gray, turquoise, or light blue. $4.77. [$125-185] **"Billings" Model Rayon Plaid.** Turquoise, beige, or gold. $6.67. [$85-95] Fall/Winter 1957.

Embroidered rayon challis, woven gingham plaid, and solid form fitted model. Turquoise, off-white, tan, black, red, or light brown. $4.90-$7.90. [$85-115] Fall/Winter 1959.

Men's "Go Western" shirts in a dressy embroidered "Golden Rider" version. Turquoise, white, or black for $7.90 [$115-135], and a form-fitting "Sharp-shooter" model in tan, turquoise, wine or black for $4.90. [$95-125] Rayon challis/gabardine. Fall/Winter 1958.

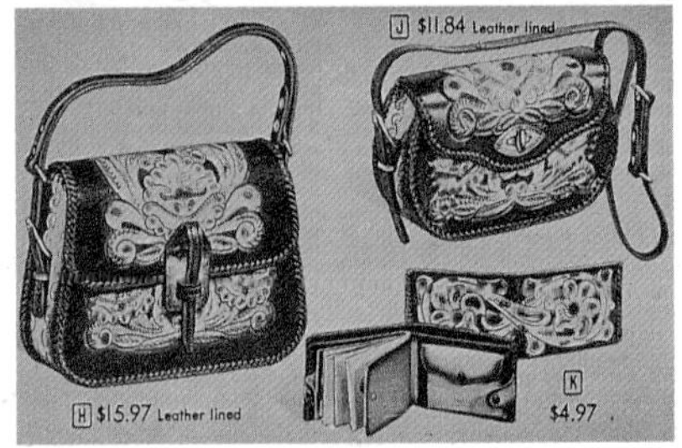

Women's hand-tooled Western top grain cowhide purses and wallet with hand laced edges and leather lining. $4.97-$15.97. [$75-125] Fall/Winter 1958.

Western wear extended to slippers and other indoor footwear. Roy Rogers leather boots and mocs were best sellers for boys and girls. $1.86-$2.83. [$55-85] Fall/Winter 1958.

Boys' Roy Rogers Dress Leather Boots. Roy's official spread-eagle emblem and monogram on each boot. Free photo of Roy with every pair purchased. Round toe or modified cowboy toe and heel, brown. $7.74. [$65-95] Fall/Winter 1957.

Fun western rainboots. Roy Rogers and Trigger pictured. Plastic, $1.87. [$65-70] Rubber, $3.77. [$75-85] Fall/Winter 1959.

Authentic Roy Rogers Biltwel Boots with rubber heels, leather-lined legs, and leather soles. Black, brown or red leather. $4.77. [$95-125] Fall/Winter 1959.

New Roy Rogers Cowboy Boots in authentic stovepipe style, fine quality leather with inlaid eagle and stars. Includes photo of Roy. Brown or black. $7.74. [$165-225] **Roy Rogers and Trigger Embossed Leather Boots.** Square or round toe, black or brown leather. $5.74. [$195-250] Fall/Winter 1959.